S0-BXE-392

HEREDITY OF THE BLOOD GROUPS

Karl Landsteiner

Father of Immunohematology

(1868–1943)

HEREDITY OF THE BLOOD GROUPS

By ALEXANDER S. WIENER, A.B., M.D., F.A.C.P, F.C.A.P.

Senior Bacteriologist (Serology) to the Office of the Chief Medical Examiner of New York City; Assistant Professor, Department of Forensic Medicine, New York University Postgraduate Medical School; Attending Immunohematologist, Jewish and Adelphi Hospitals, Brooklyn

and IRVING B. WEXLER, A.B., M.D., F.A.C.P.

Associate Pediatrician, Jewish Hospital of Brooklyn; Associate Immunohematologist, Jewish and Adelphi Hospitals, Brooklyn; Assistant Clinical Professor of Pediatrics, State University of New York, College of Medicine at New York City

GRUNE & STRATTON · *New York and London*

1958

Contents

Library of Congress Catalog Card Number 58-10790

Printed in U.S.A. (B)

Preface

DURING the past 15 years, knowledge of the blood groups has progressed with great rapidity, and blood grouping has become a fundamental biological science with applications in clinical and legal medicine, anthropology, human genetics, evolution, tissue transplantation, and veterinary medicine.

This rapid development followed the discovery of the Rh factor by Landsteiner and Wiener and the subsequent demonstration of its clinical importance in transfusion and in isoimmunization by pregnancy. When certain clinical problems could not be solved by the use of the classical agglutination techniques, new methods of blood grouping were devised which disclosed the existence of a second and more important form of antibody, namely, the so-called blocking or "univalent" antibody. These new and more sensitive techniques, the conglutination test, the anti-globulin test, and the proteolytic enzyme test became the means for recognizing instances of isoimmunization to other blood factors, so that up to present time more than 30 blood factors belonging to 9 different blood group systems have been discovered. The purpose of this monograph is to present in a unified manner the facts known about the heredity of the human blood groups.

The data that the specialist in immunogenetics must master are numerous and exceedingly complex, but fortunately the fundamental concepts are relatively simple. The reader will find that once he has mastered these few fundamental principles, he will have no difficulty in understanding and learning the detailed facts. For example, in all cases heredity is by multiple allelic genes, not only in the A-B-O system, but also in the Rh-Hr and M-N-S systems, and even in the case of the complex B blood groups of cattle and other blood group systems in animals. Therefore, emphasis is placed on principles throughout the text, and the analogies in the heredity of the various blood groups systems are pointed out. Blood grouping is primarily a subspecialty of immunology, so that for mastery of the field thorough understanding of the nature of serological reactions in general is essential. Therefore, at the very onset, the important distinction between a blood factor and agglutinogen is pointed out, and this concept is applied to explain the serology, genetics, and nomenclature of the A-B-O, Rh-Hr, M-N-S, and other blood group systems. This basic concept arises from the serological observation that antigens in general have multiple corresponding antibodies, so that there is no simple one-to-one correspondence between antigens and antibodies. The emphasis placed on this principle is unique among monographs on the subject, and, in the opinion of the authors, serves to resolve the contradictions and paradoxes which have inevitably arisen from disregard of the principle.

As Darwin has stated, false theories do little harm, while false facts may retard the progress of science. In fact, an incorrect theory may prove helpful so long as it is used only as a blueprint or working hypothesis to give a direction to investigative work. Therefore, a debt of gratitude is due to British investigators, notably R. R. Race and R. Sanger, whose observations and ideas have acted as a stimulus to investigators in the field for the past 15 years. However, if an investigator refuses to relinquish a theory refuted by readily available facts, or if he allows his preconceived ideas to color his judgment in reporting observations, serious errors may result. This has indeed occurred in the case of the Rh-Hr types, as exemplified by the erroneous reports of discovery of anti-d, reports of non-specific rises in maternal Rh antibody titers during pregnancy with Rh-negative fetuses, etc. The literature of the blood groups has been swamped with reports by poorly trained and inexperienced workers, unfamiliar with Landsteiner's and other basic work in the field. Such workers have the confidence born of ignorance, and are insensitive to the possibility of human error, so they fail to use the blind technique which is so essential for accurate results, as when carrying out the delicate tests for the Lewis blood types. Since the results which are most difficult to explain are those which are not true, an attempt has been made to include in this monograph only those reports which appear to be reliable. For any errors of omission or commission, the authors assume full responsibility, and they would appreciate having their attention called to any such errors which may come to the attention of readers.

Readers interested in other aspects of blood grouping are referred to the senior author's "Blood Groups and Transfusion," the third and last edition of which was published in 1943. That book covers the earlier literature on all aspects of blood grouping and its applications, but to do justice to the subject would now require a large volume or volumes, in view of the great growth of the field during the past 15 years. For more detailed information regarding the Rh-Hr types and their applications, the reader may consult the "Rh-Hr Types," published in 1954, a collection of the more important contributions of Wiener and his collaborators, together with a review of the literature. A simplified, compact but comprehensive account can be found in Wiener's "Rh-Hr Syllabus," which has been published also in German, Italian, Spanish, and Japanese.

The serological principles underlying blood group reactions were first pointed out by Karl Landsteiner, the father of the field of immunohematology, who received the Nobel Prize in Medicine in 1930 for the discovery of the blood groups. The senior author was fortunate to be closely associated with Karl Landsteiner from 1929 until his death in 1943. During that period, he had the advantage of Landsteiner's advice and help as mentor, collaborator, and personal friend. In appreciation of Karl Landsteiner's pathfinding contributions to the field, and as a token of personal gratitude, the authors dedicate this book to his memory on this, the 90th anniversary of his birth.

March 1958

A. S. Wiener
I. B. Wexler

CHAPTER I
Introduction

More data are available concerning blood group differences than any other heritable characteristic of healthy human beings. The reasons for this are manifold. The techniques of blood grouping have been perfected during the past two decades to a point where the blood groups of any individual can be readily and precisely determined. Moreover, individual blood group differences have no selective value, aside from the effect of isosensitization on the fetus *in utero*, so that polymorphism is the rule in almost all human populations. The blood groups are determinable at birth and remain unchanged throughout life, unaffected by age, diet, radiation or disease. Because of the implications of the blood groups in clinical medicine, for the selection of donors for blood transfusion and for isoimmunization in pregnancy, millions of blood grouping tests are performed every year. Furthermore, because every transfusion and every pregnancy is a possible source of isosensitization, new blood group antibodies are constantly being discovered and the heredity of the corresponding blood factors worked out.[1]

Principles of Blood Group Serology

For blood grouping tests, blood is obtained, preferably from a vein, and separated into its two gross components which are present in approximately equal proportions, namely, the red blood cells and the plasma in which the red blood cells are suspended within the circulation. (Serum is the fluid which is expressed from clotted blood; that is, plasma without the fibrinogen).

The red cell or erthrocyte has a lipoprotein envelope, which from its serological behavior appears to have a regularly patterned mosaic structure. Presumably certain chemical structures, named agglutinogens because of the nature of the serological tests used to demonstrate their presence, are repeated about the surface of the erythrocyte. The first discovered of these agglutinogens were the so-called A-B-O substances, which determine the four classic blood groups described by Karl Landsteiner[2] at the turn of the century. Two other sets of agglutinogens are the M-N substances and the P-p substances, both of which were discovered by Landsteiner and Levine[3] about three decades ago. Aside from the A-B-O substances, the most important are the Rh-Hr substances, discovered by Landsteiner and Wiener[4] about two decades ago. The discovery of the Rh-Hr substances led to new methods of blood typing, and as a result during the past two decades the literature has been rich with reports of the discovery of other blood group substances such a K-k (Kell), F-f (Duffy), and J-j (Kidd). Figure 1 is a hypothetical diagram of the mosaic arrangement of the blood group substances as they may be visualized on the surface of the red blood cell.

To determine an individual's blood group, a battery of diagnostic reagents or antiserums, anti-**A**, anti-**B**, anti-**M**, anti-**N**, anti-$\mathbf{Rh}_0$, anti-**rh′**, etc. are required. These reagents contain only antibodies specific for the blood factor in question, and no other antibodies. These antiserums can be obtained in a variety of ways. Anti-**A** and anti-**B** occur spontaneously in the serums of healthy adults whose red blood cells lack the corresponding agglutinogen. Since the spontaneously occurring antibodies are generally weak, donors are injected with

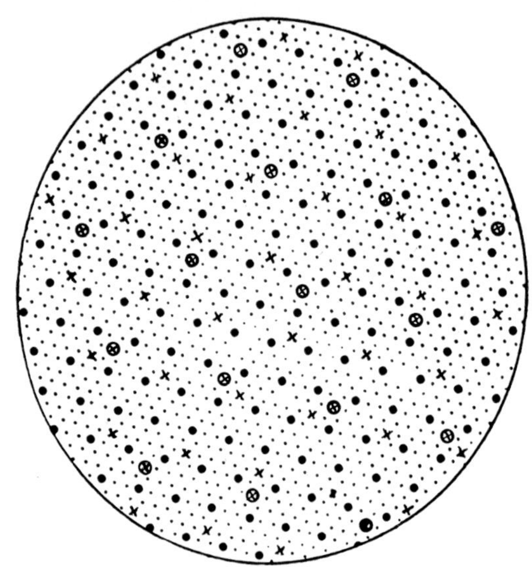

Fig. 1—Diagrammatic Representation of the Distribution of Blood Group Agglutinogen Loci on the Surface of the Erythrocyte. Legend: ● = A-B-O Haptens; × = M-N Haptens; ⊗ = Rh-Hr Haptens; • = Haptens of other specificities; human species-specific haptens.

materials of animal origin which contain A and B-like substances, and in that way antiserums of high titer and potency can be obtained. Anti-**M** and anti-**N** serums are produced by immunizing rabbits. Most of the other antiserums are obtained from mothers of erythroblastotic babies, or from patients who have had hemolytic transfusion reactions, or from deliberately immunized blood donors.[5] Recently, it has been found[6] that certain seeds contain proteins which have blood group specificity, namely, anti-**A**, anti-$\mathbf{A_1}$, anti-**B**, anti-**H**, and anti-**N**, but such reagents have not yet found wide usage. Most experts in the field prepare their own antiserums, but satisfactory reagents are also available commercially. It may be of interest to mention that, using all available reagents, as many as thirty blood factors of different specificities have been identified to date.

The blood factors detected by the use of the various antiserums are not all independent of one another, but fall naturally into sets which are known as blood group systems. Whether or not two blood factors are independent of one another or are members of a common blood group system can be determined from their distribution in the general population, by using a 2 × 2 contingency table.* For example, in the

* In order to determine whether two traits, X and Y, are independent of one another, a table is set up as follows:

Trait	X+	X−	Totals
Y+	a	b	$a + b$
Y−	c	d	$c + d$
Totals..........	$a + c$	$b + d$	N

In this table, a, b, c and d represent the frequencies of the four possible combinations of the traits X and Y, and $N = a + b + c + d$, the total number of individuals examined. If X and Y are independent, then the frequency of the Y trait among X+ individuals should be the same as its frequency amongst X− individuals, that is, $\frac{a}{a + c} = \frac{b}{b + d}$, so that $ad = bc$, or $ad - bc = 0$. In cases where $ad - bc$ is not equal to zero, whether the difference from zero is statistically significant can be determined by deriving the value of chi square with the aid of the following formula:

$$\chi^2 = \frac{(ad - bc)^2 N}{(a + b)(a + c)(b + d)(c + d)}$$

The number of degrees of freedom is taken equal to one, when determining the value of P from the table of chi square.[7]

TABLE 1

HISTORY AND PRESENT STATUS OF THE HUMAN BLOOD GROUPS

Blood Group System	Serology			Genetics		
	Blood factors and blood groups	Discoverers	Year	Nature of contribution	Investigators	Year
A-B-O and Lewis	Factors **A** and **B**	Landsteiner	1900	First family study	Epstein and Ottenberg	1908
	Blood groups O, A, and B	Landsteiner	1901	Theory of independent gene pairs	von Dungern and Hirszfeld	1910
	Blood group AB	DeCastello and Sturli	1902	Theory of multiple alleles	Bernstein	1925
	A_1-A_2 subgroups	von Dungern and Hirszfeld	1911	Extension of multiple allele theory to include subgroups A_1-A_2	Landsteiner and Levine	1927
					Thomsen, Friedenreich and Worsaae	1930
	Factor $\mathbf{F}_A$ common to human agglutinogen A and sheep blood	Schiff and Adelsberger	1924			
	Factors $\mathbf{B}_i$, $\mathbf{B}_{ii}$, $\mathbf{B}_{iii}$, etc. of agglutinogen B	Friedenreich and With	1933			
	Factor **C**, shared by agglutinogens A and B	Hooker and Anderson Landsteiner and With	1921 1926			
	Factor **H** of agglutinogens O and A_2	Schiff	1927			
	Agglutinogen A_3	Fischer and Hahn	1935	Heredity of agglutinogen A_3	Friedenreich	1936
	Agglutinogen A_4	Gammelgaard and Marcussen	1940			

	Group specific seed lectins	Renkonen	1948			
		Boyd and Reguera	1949			
	Secretor type; blood group substances in body fluids	Lehrs and Putkonen	1930	Secretor and non-secretor genes	Schiff and Sasaki	1932
	Substance **T** in non-secretor saliva	Uyeyama	1939			
	Lewis blood types	Mourant	1946	Relationship between Lewis and secretor types	Grubb	1951
		Andresen	1948			
	Blood group A_0	Dunsford and Aspinall	1952			
	Blood group A_m	Wiener	1957			
M-N-S	Factors **M** and **N** and the three M-N types	Landsteiner and Levine	1927	Heredity by pair of allelic genes	Landsteiner and Levine	1928
	Rare agglutinogen N_2	Crome	1935	Heredity of agglutinogen N_2	Friedenreich	1936
	Factors $\mathbf{M}_i$, $\mathbf{M}_{ii}$, $\mathbf{M}_{iii}$, etc.	Wiener and Landsteiner	1937			
	Factors $\mathbf{N}_i$, $\mathbf{N}_{ii}$	Wiener	1937			
	Factor **S**	Walsh and Montgomery	1947	Relationship of **S** to M-N system demonstrated	Sanger and Race	1947
	Factor **s**	Levine, Kuhmichel, Wigod, and Koch	1951			
	Factor **U**	Wiener, Unger and Gordon	1953	Heredity as simple Mendelian dominant, and probable relation to M-N types	Wiener, Unger and Cohen	1954
				Relation to M-N types proved	Greenwalt et al.	1955

TABLE 1—*Continued*

BLOOD GROUP SYSTEM	Serology			Genetics		
	Blood factors and blood groups	Discoverers	Year	Nature of contribution	Investigators	Year
	Factor **Hu**	Landsteiner, Strutton and Chase	1934	Relationship to M-N types noticed	Wiener	1936
	Factor **He**	Ikin and Mourant	1951	Relationship to M-N types demonstrated	Chalmers, Ikin and Mourant	1953
	Factor **Mi**[a]	Levine, Stock, Kuhmichel and Bronikovsky	1950	Relationship of **Mi**[a] and **Gr** to the M-N types demonstrated	Wallace et al.	1956
	Factor **Gr**	Graydon*	1946			
P	Factor **P**	Landsteiner and Levine	1927	Heredity as simple Mendelian dominant	Landsteiner and Levine	1930
	Factor **Tj**[a]	Levine, Bobbitt, Waller and Kuhmichel	1951	Relationship of **Tj**[a] to P system demonstrated	Sanger and Race	1955
Rh-Hr	First anti-rhesus (**Rh**$_0$) serum prepared; Rh-positive and Rh-negative types	Landsteiner and Wiener	1937	Heredity of **Rh**$_0$ factor as simple Mendelian dominant	Landsteiner and Wiener	1941
	Rh sensitization as cause of transfusion hemolysis	Wiener and Peters	1939			
	Rh sensitization as cause of erythroblatosis fetalis	Levine, Burnham, Katzin and Vogel	1941			

Factor **rh′**	Wiener	1941	Heredity of 4 types determined by factors **rh′** and $\mathbf{Rh_0}$, by multiple allelic genes	Wiener and Landsteiner	1941–3
Factor **hr′**	Levine and Javert	1941	Reciprocal relation of **rh′** and **hr′**,	Levine; Race and Taylor	1941–3
Factor **rh″**	Wiener and Sonn	1943	Eight Rh types determined by factors $\mathbf{Rh_0}$ **rh′** and **rh″**, and their heredity by multiple allelic genes	Wiener	1943
Intermediate Rh-Hr factors, especially $\mathfrak{R}h_0$	Wiener	1944	Extension of multiple allele theory to include Rh-Hr variants, $\mathbf{rh^w}$, **hr**, etc.	Wiener	1944–1957
Factor **hr″**	Mourant	1945	Racial distribution of the Rh-Hr types	Wiener and Landsteiner	1941–3
Factor $\mathbf{rh^w}$	Callender and Race	1946			
Factor $\mathbf{\overline{R}h_0}$	Race, Sanger and Selwyn	1950			
Factor **hr**	Rosenfield, Vogel, Gibbel	1953			
Factor $\mathbf{hr^V}$	deNatale, Cahan, Jock, Race and Sanger	1955			
Factors $\mathbf{Rh^A}$, $\mathbf{Rh^B}$, $\mathbf{Rh^C}$, etc.	Wiener and Geiger	1956			
	Rosenfield, Haber and Gibbel	1956			
	Unger and Wiener	1958			

TABLE 1—*Concluded*

Blood Group System	Serology			Genetics		
	Blood factors and blood groups	Discoverers	Year	Nature of contribution	Investigators	Year
Kell (K-k)	Factor **K** (or **Si**)	Coombs, Mourant and Race	1946	Heredity as simple Mendelian dominant	Wiener and Gordon	1947
		Wiener and Gordon	1947			
	Factor **k**	Levine, Backer, Wigod and Ponder	1949	Heredity of three K-k types by pair of allelic genes	Levine et al.	1949
	Factors **Kp**[a] and **Kp**[b]	Allen and Lewis	1957			
Duffy (F-f)	Factor **F**	Cutbush, Mollison and Parker	1950	Heredity	Cutbush and Mollison	1950
	Factor **f**	Ikin, Mourant, Pettenkofer and Blumenthal	1951			
Kidd (J-j)	Factor **J**	Allen, Diamond and Niedziela	1951	Heredity	Race, Sanger, Allen, Diamond, and Niedziela	1951
	Factor **j**	Plaut, Ikin, Mourant, Sanger and Race	1953			
Lutheran (Lu)	Factor **Lu**[a]	Callender and Race	1946	Heredity as simple Mendelian dominant	Callender and Race	1946
	Subtypes	Mainwaring and Pickles	1948			

Diego (Di)	Factor **Di**^{a}	Levine, Koch, McGee, and Hill	1954	Heredity Racial distribution	Levine et al. Layrisse and Arends	1954 1956
Unclassified	Factors of very high frequency in population					
	Vel	Sussman and Miller	1952			
	I	Wiener, Unger and Cohen	1956			
	Yt^a	Eaton et al.	1957			
	Factors of very low frequency in population					
	Levay	Callender and Race	1946			
	Jobbins	Gilbey	1947			
	Becker	Elbel and Prokop	1951			
	Ven	vanLoghem and van der Hart	1952			
	Be^a	Davidsohn, Stern, Strausser and Spurrier	1953			
	Ca	Wiener and Brancato	1953			

* Recent work of Simmons et al. has demonstrated that factor **Vw** of Hart et al. (1954) is probably identical with factor **Gr.**

case of the four blood groups O, A, B, and AB, the relationship $\bar{O} \times \overline{AB} = \bar{A} \times \bar{B}$ does not hold, which proves that factors **A** and **B** are not independent but belong to the same blood group system. The thirty-odd known blood factors have been shown to belong to some ten blood group systems, corresponding to the different chemical substances or agglutinogens that go to make up the mosaic of the surface of the red blood cell, namely, the A-B-O system, the M-N-S system, the P-p system, the Rh-Hr system, etc. (cf. table 1). A situation similar to that occurring in man has been found to exist in regard to the blood groups of cattle, fowl, dogs, mice, rats and other animals.

The tests themselves are simple, although considerable experience is required to obtain reliable and reproducible results, since the reactions are delicate and have numerous pitfalls, so that in unskilled hands they may be misinterpreted. Different methods are used depending on the reagent, but the general principle is to bring the red blood cells into contact with the appropriate antiserum and allow them to react. If the cells are agglutinated or clumped into large masses, generally visible to the naked eye, this is a positive reaction, whereas if the cells remain evenly suspended as seen under the microscope the reaction is negative (cf. fig. 2).

It is necessary to emphasize the difference between a blood factor and an agglutinogen.[8] The agglutinogen is the substance itself while the blood factors are the serological properties by which agglutinogens are recognized. Thus, when a blood is said to have the blood factor **A**, this by definition means merely that the blood cells are clumped by anti-**A** serum. It does not necessarily mean that the blood cells have the agglutinogen A of human group A blood. As a matter of fact, there are many agglutinogens other than that of the human blood group A that react with particular anti-**A** serums, notably an agglutinogen of sheep cells. Similarly, some of the blood factors **B** which characterize the agglutinogen B of human blood are shared by agglutinogens of the red cells of rabbits and other rodents.

The nature of the combination between antibody and agglutinogen can perhaps be best visualized by comparing it to the fitting of a lock by a key.[8c] The antibody is a modified serum gamma globulin which is indistinguishable by ordinary chemical means from the normal gamma globulin. It is believed that a small portion of the surface of the antibody molecule has been altered so as to make it an electro-

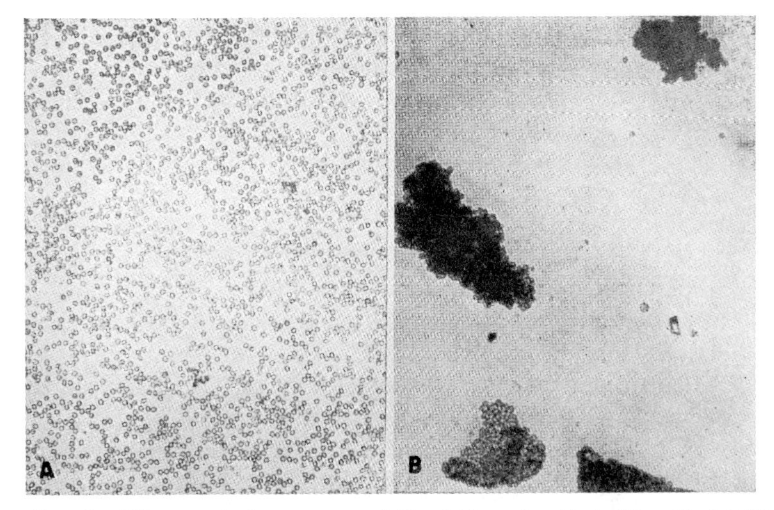

Fig. 2.—Microscopic Appearance of Blood Grouping Tests Magnified 1:80. A.: Negative Reaction (No agglutination.) B.: Positive Reaction (Showing intense agglutination of the red blood cells.)

chemical counterpart of a part of the surface of the agglutinogen molecule. In conformity with this concept it follows that not all antibodies capable of combining with an agglutinogen would necessarily have the identical configuration, any more than that all keys capable of opening a particular lock are identical. In fact, there is evidence that antiserums in general actually contain a spectrum of antibodies of related specificities, as can be demonstrated by absorption experiments. For example, anti-**M** serums prepared by immunizing rabbits by injection of human type M cells often cross react with blood cells of anthropoid apes and lower primates.[9] By absorption experiments such serum can be fractioned into five or more components, anti-$\mathbf{M}_{i}$, -$\mathbf{M}_{ii}$, -$\mathbf{M}_{iii}$, etc., each specific for a corresponding **M** factor. Thus, the human agglutinogen M is characterized not by a single blood factor but by a multiplicity of blood factors. In fact, the number of blood factors which characterize an agglutinogen is theoretically unlimited, the only practical limitation being the resourcefulness and energy of the investigator in searching for and finding new antibodies. Despite the fact that each agglutinogen has multiple blood factors, in many cases a single reagent may suffice for identification. For example, the agglutinogens A and B can be identified by the corresponding serums anti-**A** and anti-**B**, respectively. The complex agglutinogens of the Rh-Hr systems on the other hand, can be identified only by the use of a battery of antiserums.

Table 1 summarizes the history and present status of the human blood groups.

Principles of Blood Group Genetics

A-B-O Groups. The hereditary nature of the blood groups was recognized shortly after Landsteiner's discovery. The first family studies were done by Epstein and Ottenberg[10] in 1908, while in 1910 von Dungern and Hirszfeld[11] suggested that the four A-B-O groups are inherited by means of two independent pairs of allelic genes, *Aa* and *Bb*. In 1924, Felix Bernstein,[12] a mathematician, showed that this theory is incorrect, since it would require that in the general population the following relationship among the frequencies of the blood groups should hold, $\overline{\text{O}} \times \overline{\text{AB}} = \overline{\text{A}} \times \overline{\text{B}}$. As has already been shown, this relationship does not hold, which not only proves that agglutinogens A and B belong to the same blood group system, but also that their heredity is not dependent on genes located in different chromosomes. Other workers suggested that the gene pairs *Aa* and *Bb* might be linked within the same chromosome pair,[13] but as Bernstein pointed out, such a situation would yield the same result at equilibrium as independent assortment, although equilibrium is attained more slowly in the case of linkage.

According to Bernstein's theory, there are three allelic genes which have been variously designated as A, B, and O, or I^A, I^B, and I^O, respectively. (The base letter I was selected because the A-B-O blood grouping tests depend upon isoagglutination.) Accordingly, corresponding to the four blood groups there are six genotypes as follows: group O, genotype I^OI^O; group A, genotypes I^AI^A and I^AI^O; group B, genotypes I^BI^B and I^BI^O; and group AB, genotype I^AI^B. Despite Bernstein's mathematical demonstration of inheritance of the A-B-O blood groups by multiple allelic genes, the theory of linked genes enjoyed considerable vogue, until when no crossing over could be demonstrated this theory was abandoned. In 1927, Furuhata[14] in Japan proposed a theory of complete linkage, postulating the existence of three completely linked gene pairs *ab*, *Ab*, and *aB*. Obviously, this theory is identical with the theory of multiple alleles, because all that is necessary is to make the substitutions $I^A = Ab$, $I^B = aB$ and $I^O = ab$, which shows that the difference between the two concepts is merely a semantic one. Moreover, Furuhata's postulation fails to account for the absence of a linked gene pair *AB*. Therefore, Furuhata's theory is no longer mentioned in modern blood group literature. This is of interest because the concept of completely linked genes has recently been resurrected to explain the heredity of the Rh-Hr blood types and the M-N-S types.

Landsteiner and Levine[15] showed that the subgroups of group A and group AB are hereditary, and Thomsen, Friedenreich and Worsaae[16] modified Bernstein's theory to include the subgroups by postulating four allelic genes instead of three, namely, I^{A_1}, I^{A_2}, I^B, and I^O. To account for the rare agglutinogen A_3 still another allele I^{A_3} had to be invoked, and further extensions had to be made to take care of more recent discoveries.

M-N types. At first the heredity of the M-N types appeared to be simpler than the heredity of the A-B-O blood groups. To account for the existence of only three M-N types, Landsteiner and Levine[17] postulated heredity by a pair of allelic genes designated *M* and *N*, or L^M and L^N, respectively. Thus, there are three genotypes corresponding to the three phenotypes as follows: type M, genotype L^ML^M; type N, genotype L^NL^N; and type MN, genotype L^ML^N. Sanger and Race,[18] showed that the blood factor **S** discovered by Walsh and Montgomery[19] is related to the M-N system, and postulated heredity by four completely linked gene pairs as follows, *MS*, *Ms*, *NS*, *Ns*. Wiener prefers to consider this another example of multiple allelism, and has designated the four allelic genes L^S, L, l^S, and l, respectively.[20] The blood factor **U** discovered by Wiener, Unger and Gordon,[21] was shown to be related to the M-N system by Greenwalt et al.,[22] since those rare individuals whose blood cells lack the factor **U** also lack the two factor **S** and **s**. Therefore, to allow for this rare type of blood it is necessary to extend the allelic series to include at least 6 allelic genes, L^S, L^s, L^u, l^S, l^s, and l^u, respectively. To allow for still other blood factors such as **Henshaw** and **Hunter**, corresponding extensions must be made in the series of allelic genes, as shown by Shapiro.[23]

Rh-Hr Types. The most complicated example of multiple allelism in man is provided by the Rh-Hr system. When only the original $\mathbf{Rh_0}$ factor was known, Landsteiner and Wiener[24] showed that its inheritance could be explained satisfactorily by postulating a single pair of allelic genes *Rh* and *rh*. With the discovery of the two related factors **rh′** and **rh″**, Wiener[25, 26] found it necessary to postulate a minimum of 6 allelic genes r, r', r'', R^0, R^1, and R^2, respectively. With the aid of family studies and the use of anti-**hr′** serum British investigators[27] succeeded in demonstrating the rare gene R^z, to which was later added the still rarer gene r^y by Wiener.[28]

Wiener[29] showed that each of the Rh factors has variants, the most important of which are the variants of $\mathbf{Rh_0}$, and that these variations are hereditary, necessitating a further extension of the allelic series. Moreover, discovery of still other Rh-Hr factors, $\mathbf{rh^w}$, **hr″**, **hr**, $\mathbf{hr^V}$, $\mathbf{Rh^A}$, $\mathbf{Rh^B}$, etc., disclosed an ever increasing complexity of the Rh-Hr system, so that at the present time a minimum of 30 alleles must be postulated. Only in cattle has a situation been found more complex than that of the Rh-Hr system. Here Stormont, Owen and Irwin[30] have described a blood group system with more than 100 alleles.

Probably, the other blood group systems in man similarly depend upon multiple allelic genes for their hereditary transmission. For example, the P blood types which were originally considered to be inherited by a pair of alleles, *P* and *p*, are now known to be more complex, since Sanger[31] showed that the rare $\mathbf{Tj^a}$-negative type described by Levine[32] depends on a corresponding rare allelic gene belonging to the P system. Moreover, the recent observations of Allen[33] show that the K-k system is also inherited by a complex series of multiple alleles. The same probably applies to the other blood group systems, as will be shown when these are discussed individually in detail.

None of the blood group genes is known to be sex linked, and the genes of each blood group system appears to be located on different pairs of chromosomes, as is shown by their independent heredity. Thus, each blood group system can be used as a marker for studies in human linkage. Other applications of the hereditary nature of the blood group systems are as a test for the zygosity of twins,[249] and in medicolegal problems of disputed parentage.

Nomenclature

When devising symbols for the blood groups one must take into account the distinction between agglutinogens and blood factors, and between genotypes and phenotypes. Therefore, the convention has been adopted of using regular type for symbols for agglutinogens and phenotypes, bold-face type for symbols representing blood factors and their corresponding antibodies, while italics are used for the symbols representing genes and genotypes. For example, blood group A is characterized by the presence in the red cells of agglutinogen A as evidenced by clumping of the red cells by anti-**A** but not by anti-**B** serum, which shows the presence in the cells of blood factor **A**, but not blood factor **B**. Moreover, agglutinogen A is inherited by means of a corresponding allelic gene I^A, so that corresponding to group A there are two theoretically possible genotypes, I^AI^A and I^AI^O. The full implications of this terminology will become clearer as each individual blood group system is discussed.

CHAPTER II

The A-B-O Blood Groups and Lewis Types

Landsteiner's A-B-O Groups

Serology. According to Landsteiner's original theory, the four blood groups are determined by two agglutinogens A and B on the red blood cells and two corresponding naturally occurring iso-agglutinins in the serum, anti-**A** (alpha) and anti-**B** (beta) respectively. Thus, the complete serological formula of the four blood groups is as follows: O$\alpha\beta$, Aβ, Bα, and AB*o*. According to Landsteiner's law the alpha and beta agglutinins are regularly present in the serum when the corresponding agglutinogen is absent from the red cells, except during the neonatal period, when the antibody producing mechanism is immature.

Gene Frequency Analysis. Bernstein's theory of the heredity of the A-B-O blood groups has already been described. The accuracy of the theory has been confirmed by family studies, and by statistical studies on the distribution of the blood groups in the general population. Bernstein's method of gene frequency analysis is as follows. Let p, q, and r represent the frequency of the three allelic genes I^A, I^B and I^O, respectively, so that $p + q + r = 1$, or 100 per cent. Then the frequencies of the four phenotypes can be expressed as follows:

Phenotypes	Genotypes	Frequencies
O	I^OI^O	r^2
A	I^AI^A and I^AI^O	$p^2 + 2pr$
B	I^BI^B and I^BI^O	$q^2 + 2qr$
AB	I^AI^B	$2pq$

Thus, if $\bar{\text{O}}$, $\bar{\text{A}}$, $\bar{\text{B}}$, and $\overline{\text{AB}}$ represent the frequencies of the four blood groups in the general population, then $r^2 = \bar{\text{O}}$

$$\text{and } r = \sqrt{\bar{O}} \tag{1}$$

$$\text{Moreover, } p^2 + 2pr = \bar{A}$$

$$\text{so that } p^2 + 2pr + r^2 = (p + r)^2 = \bar{O} + \bar{A}$$

$$\text{Therefore, } p + r = \sqrt{\bar{O} + \bar{A}}$$

$$\text{so that } p = \sqrt{\bar{O} + \bar{A}} - \sqrt{\bar{O}} \tag{2}$$

$$\text{Similarly, } q = \sqrt{\bar{O} + \bar{B}} - \sqrt{\bar{O}} \tag{3}$$

$$\text{Since, } p + q + r = 1$$

$$\text{then } \sqrt{\bar{O} + \bar{A}} + \sqrt{\bar{O} + \bar{B}} - \sqrt{\bar{O}} = 1 \tag{4}$$

By substituting the frequencies of the blood groups A, B, and O of any population in formula (4) above, the validity of the theory of multiple alleles can be tested. Application of this test to large series of populations has served to substantiate Bernstein's theory. The test is only valid with populations at genetic equilibrium, that is, when there is random mating or panmixia. It can readily be shown that after a single generation of random matings the population will reach such equilibrium.[34]

Actually, the formulae used by Bernstein himself in making this test are slightly different from the ones given above. Since $p + q + r = 1$, then the following set of formulae may be derived from formulae (1), (2), and (3) above,

$$r = \sqrt{\bar{O}} \tag{5}$$

$$p = 1 - \sqrt{\bar{O} + \bar{B}} \tag{6}$$

$$q = 1 - \sqrt{\bar{O} + \bar{A}} \tag{7}$$

In making the statistical test, Bernstein derived the values of p, q, and r from the formulae (5) to (7), and then calculated the deviation D of the sum from 100 per cent.

Thus,

$$D = 1 - (p + q + r) \tag{8}$$

To test whether the value of the deviation D is significant for a particular population, the standard error of D must be known. This Bernstein showed to be as follows:

$$\sigma_D = \sqrt{\frac{pq}{2N(1 - p)(1 - q)}} \tag{9}$$

where N represents the number of individuals in the sample of the population being examined, and p and q are the calculated frequencies of genes I^A and I^B, respectively.

For the purpose of testing the validity of the theory of multiple alleles, the original report of L. and H. Hirszfeld,[35] who were the first to demonstrate the differences in the distribution of the A-B-O blood groups in different ethnic groups, is ideal. The data of these authors were collected during the first world war in the Balkans, six years before Bernstein proposed his genetic theory. The observations of L. and H. Hirszfeld and the results of Bernstein's statistical test of the theory of multiple alleles are given in table 2. It can readily be seen that results conform satisfactorily with the expectations under the theory.

In general, the calculated values of p, q and r would not be expected to add up to 100 per cent exactly. The values of these genes frequencies, as calculated by either set of formulae, while consistent, do not have the maximum efficiency in the statistical sense. The most efficient set of estimates can be obtained by the maximum likelihood method, or more simply by the formulae suggested by Bernstein,[36] as follows:

$$p' = p\left(1 + \frac{D}{2}\right) \tag{10}$$

$$q' = q\left(1 + \frac{D}{2}\right) \tag{11}$$

$$r' = \left(r + \frac{D}{2}\right)\left(1 + \frac{D}{2}\right) \tag{12}$$

where p, q, and r, are the first estimates of the gene frequencies obtained with formulae (5) to (7), while p', q' and r' are the improved estimates. These formulae are of particular value to anthropologists who are interested in population studies on the A-B-O blood groups. The sum of the values, p', q', and r' will not equal 100 per cent exactly, but $1 - \frac{D^2}{4}$. Since D generally will be small, the difference $\frac{D^2}{4}$ will be insignificant.

Family studies. For a satisfactory test of the theory of multiple alleles by family studies, the fact that group A and group B indi-

TABLE 2

STATISTICAL TEST OF BERNSTEIN'S THEORY

(Made with data accumulated by L. and H. Hirszfeld)

Race	Number of People	Frequencies of Groups				p	q	r	$p+q+r$	Dev.	P.E.	$\frac{\text{Dev.}}{\text{P.E.}}$
		O	A	B	AB							
English	500	0.464	0.434	0.072	0.031	0.268	0.052	0.681	1.001	0.001	0.0037	0.3
French	500	0.432	0.426	0.112	0.030	0.262	0.074	0.657	0.993	0.007	0.0037	1.9
Italians	500	0.472	0.380	0.110	0.038	0.237	0.077	0.687	1.001	0.001	0.0034	0.3
Serbians	500	0.380	0.418	0.156	0.046	0.268	0.107	0.516	0.991	0.009	0.0045	2.0
Greeks	500	0.382	0.416	0.162	0.040	0.262	0.107	0.618	0.987	0.013	0.0044	3.0
Bulgarians	500	0.390	0.406	0.142	0.062	0.271	0.108	0.624	1.003	0.003	0.0045	0.7
Arabs	500	0.432	0.324	0.190	0.050	0.209	0.129	0.660	0.998	0.002	0.0042	0.5
Turks (Macedonia)	500	0.368	0.380	0.186	0.066	0.256	0.136	0.607	0.999	0.001	0.0049	0.2
Russians	1000	0.407	0.312	0.218	0.063	0.210	0.152	0.638	1.000	0	0.0030	0
Spanish Jews	500	0.388	0.330	0.232	0.050	0.213	0.153	0.623	0.989	0.011	0.0047	2.3
Madagascans	400	0.458	0.262	0.237	0.045	0.168	0.154	0.675	0.997	0.003	0.0046	0.7
Senegal Negroes	500	0.432	0.224	0.292	0.050	0.149	0.189	0.657	0.995	0.005	0.0043	1.2
Annamese	500	0.420	0.224	0.284	0.072	0.161	0.198	0.648	1.008	0.008	0.0045	1.8
Hindus	1000	0.313	0.190	0.412	0.085	0.149	0.291	0.560	1.000	0	0.0040	0

After Wiener: *Blood Groups and Transfusion*, 3rd ed., C. C Thomas, Springfield, Ill. (1943).

viduals may be homozygous or heterozygous must be taken into account. If the distribution of the blood groups in the general population is known, the proportion of homozygous and heterozygous individuals can be estimated from the frequency of the genotypes as follows:

$$\text{Genotype } I^A I^A = p^2$$
$$\text{and genotype } I^A I^O = 2pr$$

Therefore,

$$\text{the proportion of homozygous individuals in group A} = \frac{p^2}{p^2 + 2pr} \quad (13)$$

Similarly,

$$\text{the proportion of homozygous individuals in group B} = \frac{q^2}{q^2 + 2qr} \quad (14)$$

When making these calculations, it is preferable to use the improved estimates of the gene frequencies given by formulae (10) to (12).

It is frequently possible to distinguish homozygous and heterozygous individuals by studying the blood groups of parents and children. For example, if a group A individual has a group O parent or child, he must of necessity be of genotype $I^A I^O$. When neither of the parents and none of the children are group O this leaves open the question of zygosity, except in certain special cases such as when both parents are group AB.

Sometimes, it is possible to distinguish heterozygous from homozygous individuals by serological means, since individuals homozygous for a particular blood factor generally have red blood cells which are clumped more strongly by the corresponding antiserums. This is known as the gene dose effect. The gene dose effect, however, is not sufficiently pronounced to be dependable in the case of the A-B-O groups, and, in addition, the existence of the subgroups of group A interferes with the interpretation of the reactions. The possibility that an anti-**O** reagent might be prepared which could detect the presence of a product of gene I^O has frequently been discussed in the literature. Such antibodies have indeed been found in the serum of certain sensitized group AB persons,[37, 38] but they are extremely rare and those encountered to date have been of too low titer and avidity to yield reliable results.

If the gene frequencies p, q, and r for the general population are known, the percentages of children of the various types for eveyr possible mating can be calculated. Schiff[39] has derived general formulae for expressing these frequencies in terms of p, q and r, as shown in table 3. From the qualitative point of view, it is evident that no child will have blood containing the agglutinogen A or B unless that agglutinogen is present in the blood of one or both parents. In addition, a parent of group AB cannot have a group O child, and a group O parent cannot have a child of group AB. Studies on thousands of families by many independent workers have yielded data supporting the expectations under the theory of multiple alleles, when allowance is made for technical errors and a certain incidence of illegitimacy.

The study of mother-child combinations can also yield useful data for testing the genetic theory. There are 16 different theoretically possible mother-child combinations, but of these two are excluded under the theory of multiple alleles, namely, mother AB and child O, and group O mother with an AB child. In such studies the problem of illegitimacy does not enter, and it is significant that the results show no deviation from the genetic theory. Formulae for the 14 possible mother-child combinations in terms of p, q and r are readily derived as shown in table 4.

Subgroups of A

Gene Frequency Analysis. As already mentioned the two most common subgroups of group A and group AB are readily included in the scheme by postulating two allelic genes, I^{A_1} and I^{A_2}, in place of I^A, where I^{A_1} is dominant to I^{A_2}. Corresponding to the 6 possible phenotypes there are 10 possible genotypes as follows: group O: genotype I^OI^O; subgroup A_1: genotypes $I^{A_1}I^{A_1}$, $I^{A_1}I^{A_2}$, and $I^{A_1}I^O$; subgroup A_2: genotypes $I^{A_2}I^{A_2}$ and $I^{A_2}I^O$; group B: genotypes I^BI^B and I^BI^O; subgroup A_1B: genotype $I^{A_1}I^B$; and subgroup A_2B: genotype $I^{A_2}I^B$. If the frequencies of the 4 alleles I^{A_1}, I^{A_2}, I^B, and I^O are represented by the symbols p_1, p_2, q and r, respectively, then the frequencies of the genes can be calculated from the frequencies of the 6 phenotypes in the population, with the aid of the following formulae, which are derived like the formulae for p, q and r of equations (1) to (3).

$$p_1 = \sqrt{\bar{O} + \bar{A}_1 + \bar{A}_2} - \sqrt{\bar{O} + \bar{A}_2} \qquad (15)$$

$$p_2 = \sqrt{\bar{O} + \bar{A}_2} - \sqrt{\bar{O}} \qquad (16)$$

$$q = \sqrt{\bar{O} + \bar{B}} - \sqrt{\bar{O}} \qquad (17)$$

$$r = \sqrt{\bar{O}} \qquad (18)$$

TABLE 3

FREQUENCIES OF TYPES OF MATINGS AND THEIR OFFSPRING

Type of Mating	Frequency of Mating	Offspring			
		O	A	B	AB
O × O	r^4	r^4	none	none	none
O × A	$2pr^2(p + 2r)$	$2pr^3$	$2pr^2(p + r)$	none	none
O × B	$2qr^2(q + 2r)$	$2qr^3$	none	$2qr^2(q + r)$	none
O × AB	$4pqr^2$	none	$2pqr^2$	$2pqr^2$	none
A × A	$p^2(p + 2r)^2$	p^2r^2	$p^2(p + r)(p + 3r)$	none	none
A × B	$2pq(p + 2r)(q + 2r)$	$2pqr^2$	$2pqr(p + r)$	$2pqr(q + r)$	$2pq(p + r)(q + r)$
B × B	$q^2(q + 2r)^2$	q^2r^2	none	$q^2(q + r)(q + 3r)$	none
A × AB	$4p^2q(p + 2r)$	none	$2p^2q(p + 2r)$	$2p^2qr$	$2p^2q(p + r)$
B × AB	$4pq^2(q + 2r)$	none	$2pq^2r$	$2pq^2(q + 2r)$	$2pq^2(q + r)$
AB × AB	$4p^2q^2$	none	p^2q^2	p^2q^2	$2p^2q^2$
Totals.........	1.00	r^2	$p^2 + 2pr$	$q^2 + 2qr$	$2pq$

Modified after Schiff: *Technik der Blutgruppenuntersuchung*, p. 68, Julius Springer, Berlin (1932).

TABLE 4
FREQUENCIES OF MOTHER-CHILD COMBINATIONS IN TERMS OF p, q, AND r
(After Schiff)*

Group of Mother	Children's Groups			
	O	A	B	AB
O†	r^3	pr^2	qr^2	0
A	pr^2	$p(p^2 + 3pr + r^2)$	pqr	$pq(p + r)$
B	qr^2	pqr	$q(q^2 + 3qr + r^2)$	$pq(q + r)$
AB	0	$pq(p + r)$	$pq(q + r)$	$pq(p + q)$

* Schiff, *Klin. Woch.* **6:** 303 (1927).
† As can be seen, if families with group O mothers are considered separately, the percentages of children of groups O, A, and B are equal to the gene frequencies, r, p, and q, respectively.

The values for p_1, p_2, q and r, obtained by formulae (15) to (18), while consistent, are not efficient, and generally their sum is not exactly equal to 100 per cent. A corrected set of gene frequencies which closely approximates the maximum likelihood estimates, can readily be obtained by starting with the values of p', q' and r' given by formulae (10) to (12), and then taking

$$p_1' = \frac{p' \cdot p_1}{p_1 + p_2} \tag{19}$$

$$p_2' = \frac{p' \cdot p_2}{p_1 + p_2} \tag{20}$$

Serology. As has already been mentioned the agglutinogen A occurs in two principal forms, designated as A· and A_2, respectively. Anti-**A** serum, which is generally obtained from group B individuals, contains a mixture of antibodies of closely related specificities. If such a serum is absorbed with red cells of subgroup A_2, a reagent can be obtained which clumps A_1 cells but not A_2 cells. This reagent, known as anti-$\mathbf{A_1}$ serum, can be used to subdivide group A individuals into subgroups A_1 and A_2, and group AB individuals into subgroups A_1B and A_2B, respectively.

There is evidence that the difference between A_1 and A_2 is qualitative, yet the serological reactions disclose certain striking quantitative effects. Absorption of anti-**A** serum (from group B individuals), for example, by a small amount of A_1 cells has almost the same effect as absorption by a larger amount of A_2 cells. In fact, when the anti-$\mathbf{A_1}$ reagent is prepared, if an excess of A_2 cells is used for the absorption the reactivity for A_1 cells may be destroyed. Thus, the difference between agglutinogens A_1 and A_2 is less sharp than between agglutinogens A and B. Another difficulty is that the presence of agglutinogen B reduces the activity of the agglutinogen A and vice versa, if both agglutinogens are present in the same red cell. This is particularly noticeable in the case of blood of subgroup A_2B, where the reactivity of the A agglutinogen is so diminished that its presence may be overlooked unless potent anti-**A** reagents are used. Accordingly, a common error is to classify individuals of subgroup A_2B incorrectly as group B.

Another difficulty in technique is created by the fact that in newborn babies the red cells are less agglutinable by anti-**A** and anti-**B** sera, and as a result red cells from newborn infants of subgroup A_1 may fail to react with anti-$\mathbf{A_1}$ serum and be classified incorrectly as subgroup A_2. As may be anticipated, the difficulty in determining the subgroups is considerably increased when the baby belongs to group AB. Not infrequently, therefore, it is necessary to wait until the infant is 6 to 12 months old before the subgroups can be reliably determined. Still another difficulty results from the existence of blood cells which give reactions intermediate between A_1 and A_2; this intermediate subgroup occurs most frequently among Negroids. While anti-$\mathbf{A_1}$ lectins (reagents prepared from seeds) may give more striking reactions than anti-$\mathbf{A_1}$ reagents prepared from human group B serum, their use does not obviate the difficulties encountered in subgrouping group A and group AB blood, especially of newborn infants and Negroids.

A useful reagent to have when carrying out subgrouping tests is the so-called anti-**O** (A_2) or anti-**H** serum. This reagent reacts regularly with red cells of group O and of subgroup A_2, and more weakly with cells of the other groups. Landsteiner and Levine[40] first suggested the use of such reagents to confirm the results of the subgrouping tests.

Family Studies. The theory of inheritance of the subgroups of A by means of the corresponding allelic genes I^{A_1} and I^{A_2} has been verified by studies in families by many independent investigators. The studies carried out by Wiener and his collaborators between the years 1930 and 1941 are summarized in table 5. According to the theory, the following rules of heredity must hold.

TABLE 5

SUMMARY OF PUBLISHED STUDIES ON THE HEREDITY OF THE SUBGROUPS OF GROUP A AND GROUP B

(After Wiener, *Blood Groups and Transfusion*, 3rd edition, 1943)

Parental Combinations	Number of Families	Number of Children in each Group						
		O	A_1	A_2	B	A_1B	A_2B	Totals
$A_2 \times O$	105	124	(4)	163	0	0	0	291
$A_2 \times A_2$	13	13	(1)	26	0	0	0	40
$A_2 \times B$	43	21	0	34	29	0	32	116
$A_2B \times O$	16	0	0	37	27	0	0	64
$A_2B \times A_2$	3	0	0	3	3	0	1	7
$A_2B \times B$	10	(1)	0	10	15	0	6	32
$A_1 \times O$	387	404	695	69	0	(1)	0	1169
$A_1 \times A_1$	168	79	394	16	0	0	(1)	490
$A_1 \times A_2$	79	52	127	54	0	0	0	233
$A_1 \times B$	120	57	90	8	54	84	6	299
$A_1 \times A_2B$	16	0	14	9	10	8	2	43
$A_1B \times O$	38	0	55	0	53	0	0	108
$A_1B \times A_1$	35	0	65	0	19	40	5	129
$A_1B \times A_2$	8	0	18	0	5	(1)	10	34
$A_1B \times B$	25	0	15	0	32	16	0	63
$A_1B \times A_1B$	2	0	6	0	2	8	0	16
Totals.........	1,068	751	1,484	429	249	158	63	3,134

Note: The rare parental combination, $A_1B \times A_2B$ and $A_2B \times A_2B$ were not encountered in these studies.

1). Agglutinogen A_1 cannot occur in a child's blood unless it is present in one or both parents.

2). The combination of an A_1B parent with an A_2 child, or an A_2 parent with an A_1B child cannot occur.

3). In the mating $A_1 \times O$, if there are any children of group O there cannot be any children of group A_2, and, conversely, if there are any children of subgroup A_2 there cannot be any children of group O. (There are other matings in which similar rules can be enunciated, depending upon the fact that an individual of subgroup A_1 must belong to one of the three genotypes, $I^{A_1}I^{A_1}$, $I^{A_1}I^{A_2}$, or $I^{A_1}I^{O}$.)

4). In the matings $A_1B \times A_1B$ and $A_1B \times B$, children of subgroup A_2B cannot occur.

Among the 3,134 children of table 5, there were a number of apparent exceptions to the genetic theory. These, which are indicated by parentheses, are present in a larger number than may reasonably be attributed to illegitimacy alone. Probably an important cause for these discrepancies is errors in classifying individuals, due to the difficulties of subgrouping already pointed out.

Subgroups A_3 and A_3B. The subgroup A_3 is a rare variant of group A which is characterized by the very weak agglutinability of the cells, even with potent anti-**A** serum.[41] Such blood is readily recognized when tested with anti-**A** serum because the clumping though distinct is incomplete, and the microscopic pattern is that of clumps of red cells on a background of many free cells.[42] Because of the low frequency of this subgroup, the number of heredity studies have been limited, but the data obtained support the theory[43] that the agglutinogen A_3 is determined by a corresponding gene I^{A_3} "recessive" to I^{A_1} and I^{A_2}, but "dominant" over I^O.

Other Subgroups. Even rarer than bloods of subgroup A_3 are those variants of agglutinogen A which are so weakly reactive, that clumping of the red cells occurs only with anti-**A** serums of extremely high titer and avidity, whether these serums are derived from group B and group O human beings, or produced by immunizing rabbits or other animals. Such bloods have been assigned the designations subgroup A_4, A_5, etc., and should not be confused with blood of group A_0 described on page 27. One may anticipate that blood of the subgroups A_3B or A_4B will in most cases be mistaken for group B blood, while in newborn babies blood of subgroups A_3 and A_4 will be practically indistinguishable from group O blood. Tests for the blood group substances in saliva (cf. page 30) will be of little or no help in resolving such problems, because of the corresponding weak inhibition reactions which will be obtained. Fortunately, in view of the extreme rarity of the agglutinogens A_3 and A_4, this problem hardly ever arises in clinical work, and has not yet arisen in any medicolegal case.

Factor *C*

When group O individuals are injected with group A red cells, not infrequently the titer of the serum is increased not only for group A cells but also for group B cells. Similarly, injection of group B red cells into group O individuals may raise the serum's titer for group A as well as for group B cells. Conversely, if group O serum is absorbed with group A cells its reactivity for group B cells may be markedly diminished. Similarly, when group O serum is absorbed with group B cells, a diminution in the reactivity against group A cells may occur. This indicates the presence in group O serum of an antibody which reacts with both group A and group B cells. This antibody has been designated as anti-**C** and the corresponding blood factor, which is common to the two agglutinogens A and B, is known as blood factor **C**.[44] Thus, in place of the simple scheme of the four blood groups, the more complex one given in table 6 is necessary to account for these observations. It may be remarked that reagents of specificity anti-**C** can also be obtained from animal serums and from seed extracts.

It is clear from table 6 that the factor **C** is another serological attribute of the agglutinogens A and B. Tests for factor **C** are not needed to *identify* the agglutinogens A and B, and need not be taken into account in heredity studies or discussions of genetic theory. The existence of factor **C** is mentioned here because of its theoretical significance in relation to the difference between a blood factor and an agglutinogen, and because of its practical importance in the pathogenesis of ABO hemolytic disease.

A subgroup of group A has been described[45] which has been variously designated as A_x or A_0. The symbol A_0 was selected by Levine[46]

TABLE 6

SEROLOGIC REACTIONS OF THE FOUR A-B-O BLOOD GROUPS

Group	Red Cells		Agglutinins in Serum†
	Agglutinogens	Blood factors	
O	—*	—*	Anti-**A**, anti-**B** and anti-**C**
A	A	**A** and **C**	Anti-**B**
B	B	**B** and **C**	Anti-**A**
AB	A and B	**A**, **B** and **C**	—

* For simplicity, agglutinogen O and blood factor **H** are omitted from this table.

† The antibodies of each of the designated specificities are not homogeneous, but consist of a "spectrum" of antibodies. Thus, in addition to anti-**A** which acts almost equally on agglutinogens A_1 and A_2, group O and group B serums contain anti-$\mathbf{A}_1$, which acts much more strongly on A_1 than on A_2. Similarly, in addition to anti-**C**, group O serum may contain anti-$\mathbf{C}_A$, reacting more strongly with agglutinogen A than agglutinogen B, and anti-$\mathbf{C}_B$, reacting more strongly with agglutinogen B than agglutinogen A.[48] Similarly, anti-**B** actually comprises a spectrum of antibodies, anti-$\mathbf{B}_i$, anti-$\mathbf{B}_{ii}$, etc.

because while such cells generally give little or no reaction with potent anti-**A** serum they are clumped by the great majority of group O sera. A more appropriate symbol for this blood group may be C, because the most prominent feature of the red cells is the presence of blood factor **C**. Moreover, the sera of such individuals usually agglutinate group A as well as group B cells, but the reactions with group A cells are generally of low titer.[47] While there is some serological evidence that the red cells may contain a very weakly reacting **A** but no **B** factor, the almost regular reactions with group O serum suggest that the most important attribute of these cells may be the presence of the **C** factor.[44] The report that group A saliva inhibits the agglutination of A_0 cells by group O serum, but not group B saliva, suggests that the factor in such cells should perhaps be designated $\mathbf{C}_A$. Heredity studies have been hampered by the very low frequency of this blood type. The irregular expression of this blood type A_0 in the few families studied suggests the operation of a modifier rather than a new allelic gene.

Serological Complexity of the A-B-O Agglutinogens

Immunization of rabbits with human group A blood may stimulate the production of antibodies which react not only with group A cells, but which also lyse sheep cells. Similarly, the injection of sheep red cells into rabbits may stimulate the production of high titer anti-**A** agglutinins. This demonstrates the existence of a blood factor common to sheep red cells and human group A blood.[49] Because of the similarity of behavior to Forssman antiserums, which also hemolyze sheep cells, the factor has been designated $\mathbf{F}_A$. When human anti-**A** serum is absorbed with sheep red cells in order to remove the $\mathbf{F}_A$ antibody, a fraction of antibody often remains which agglutinates human group A cells but does not react with sheep cells. This shows that there exist also blood factors peculiar to group A cells which are not shared with sheep red cells.

Similarly, the anti-**B** agglutinins of human group A serums are not homogeneous, as can be shown by their cross reactions with blood cells from lower animals. Thus, almost all human anti-**B** serums strongly agglutinate blood from rabbits. Absorption of such serums by rabbit red cells generally leaves behind a fraction which clumps human B cells but not rabbit cells, though some human anti-**B** serums exist which can be completely absorbed by rabbit cells as well as by

TABLE 7
SOME OF THE SEROLOGICAL PROPERTIES OF THE A-B-O AGGLUTINOGENS

Genes	Corresponding Agglutinogens	Blood Factors
I^O	O	**H**
I^{A_1}	A_1	$\mathbf{A_1}$, **A**, $\mathbf{F_A}$, **C**
I^{A_2}	A_2	**A**, **H**, $\mathbf{F_A}$, **C**
I^B	B	**B**, $\mathbf{B_{ii}}$, $\mathbf{B_{iii}}$, ... **C**

human group B cells.[50] If we designate the human anti-**B** agglutinins which react exclusively with human B cells as anti-$\mathbf{B_i}$, and the agglutinins which react with both rabbit and human B cells as anti-$\mathbf{B_{ii}}$, then it is clear that human group B cells have at least two factors $\mathbf{B_i}$ and $\mathbf{B_{ii}}$, while rabbit red cells have factor $\mathbf{B_{ii}}$ but not $\mathbf{B_i}$. Similarly, comparative studies carried out with blood of other mammals,[51] such as opposums and guinea pigs disclose additional complexities so that in general human group A serum appears to contain a spectrum of anti-**B** agglutinins while agglutinogen B has corresponding multiple blood factors $\mathbf{B_i}$, $\mathbf{B_{ii}}$, $\mathbf{B_{iii}}$, . . . etc.

The agglutinogens A and B determined by the corresponding genes I^A and I^B regularly possess *all* of the appropriate blood factors as described above, which confirms the concept that these factors are merely serological attributes of their single respective agglutinogen (cf. table 7).

Group-Specific Substances in Body Fluids and Secretions

Chemical studies have disclosed that the A-B-O blood group substances occur in the body in two forms, one soluble in organic solvents and the other as water soluble mucopolysaccharides. The former are found in the red cells of all individuals of the appropriate blood group and also in the body cells. The water soluble form does not occur in all individuals. Among Caucasoids approximately 75 per cent of all individuals (secretors) have the water soluble form in their bodies whereas 25 per cent (non-secretors) lack it or have only insignificant amounts.[52, 53]

The water-soluble group specific mucopolysaccharides are heat stable so that they can be extracted from tissues with boiling water. They occur in highest concentration in secretory organs, such as the salivary glands, gastric and duodenal mucosa, the pancreas, and the seminal vesicles. In the living subject the secretor/non-secretor status can be diagnosed most simply by testing the saliva, where the group specific polysaccharides occur in high concentration in the case of secretors.

Technique. The saliva specimens are collected in wide mouth dry tubes. Only small amounts are necessary for the test. Immediately following the collection the tubes are placed in boiling water for ten minutes to destroy the enzymes which saliva contains, since these may destroy the blood group substances. Coagulated mucus is removed by centrifugation and the supernatant opalescent fluid is used for the test. In infants and small children the saliva may be collected on cotton swabs which are expressed into narrow tubes.[54]

The principle of the test is to determine the ability of the saliva to inhibit group specific agglutination reactions. Thus, group A secretor saliva prevents anti-**A** serum from clumping group A cells, but does not prevent anti-**B** serum from clumping B cells, and similarly for group B and group AB saliva from secretors (cf. table 8). On the other hand, group O saliva and non-secretor saliva will not inhibit anti-**A** and anti-**B** serums.

Schiff has shown that saliva from secretors of all groups inhibit the agglutination of group O cells by so-called anti-**O** serums obtained from animals. Such reagents, which are now commonly designated as anti-**H** are more conveniently obtained from seed extracts, such as of *Ulex europaeus*, and these reagents are very useful for determining the secretor status of group O individuals, as well as individuals of the other three blood groups (cf. table 8). By family studies, Schiff

TABLE 8

INHIBITION TESTS FOR THE PRESENCE OF A-B-O GROUPS IN SALIVA*

Saliva of	Anti-**A** Serum + Group A_2 Cells + Saliva	Anti-**B** Serum + Group B Cells + Saliva	Anti-**H** Serum + Group O Cells + Saliva
Secretor group O	+	+	−
Secretor group A	−	+	−
Secretor group B	+	−	−
Secretor group AB	−	−	−
Non-secretor, all groups	+	+	+

* + = agglutination; − = no agglutination.

and Sasaki[55] were able to show that the secretor status of an individual is genetically determined, and depends on a pair of allelic genes *Se* and *se*. Thus secretors may be homozygous, genotype *SeSe*, or heterozygous, genotype *Sese*, while non-secretors are always homozygous, genotype *sese*. In family studies the genes *Se* and *se* segregate independently of the genes I^A, I^B, and I^O, so that they may be assumed to be located on a different pair of chromosomes. The secretor type thus provided the first example in blood group genetics of gene interaction.

Aside from the inhibition technique, the blood group substances can be demonstrated in the saliva by precipitin techniques. Thus, if a potent anti-**A** serum, prepared by immunizing rabbits, is layered with group A or group AB secretor saliva, a white ring will appear at the interface, but no such ring appears with non-secretor saliva or saliva of group O or group B. This technique is not suitable for routine use because of the difficulty of preparing the reagents. Moreover, so far attempts to produce comparable anti-**B** reagents have been unsuccessful. Of interest in this connection is the observation of Uyeyama,[56] who found that about 16 per cent of normal fowl serums contain precipitins for non-secretor saliva but not for secretor saliva. Uyeyama designated the antibody in question as anti-**T** and asserted that it detected a substance, T, present in saliva of non-secretors but not in secretors. This finding conforms with the observation that saliva from non-secretors is not devoid of mucopolysaccharides, and suggests that the non-secretor status is due to some molecular alteration in the blood group mucopolysaccharide which eliminates A-B-O activity and substitutes T reactivity. Under this concept, H and T may be considered to be contrasting products of the genes *Se* and *se*, respectively.

Lewis Blood Groups

In 1946, Mourant[57] encountered a serum in a blood donor, which agglutinated the blood cells of 24 of 96 random Englishmen of group O. The reactions of this serum was stronger at low temperature than at body temperature, the antibodies having properties of type-specific cold agglutinins as described by Landsteiner and Levine. The blood factor detected by this serum was named **Lewis** after one of the donors whose serum contained the antibody. In 1948, Andresen[58] reported that in the course of a year he and Freiesleben had encountered 8 examples of an antiserum which agglutinated the red cells of 21 per cent of adult Danes. The blood factor detected by these serums Andresen designated as **L**, which is fortunate since it proved to be identical with the **Lewis** factor of Mourant. Shortly thereafter, Andresen[59] reported the existence of a second type of serum which gave reactions almost antithetical to his original anti-**L** serum and which he now called anti-$\mathbf{L_1}$ to distinguish it from the second serum which he called anti-$\mathbf{L_2}$. In this review, however, the symbols $\mathbf{Le^a}$ and $\mathbf{Le^b}$ will be used for the two Lewis factors in conformity with the recommendation of British workers.

The facts concerning the **Lewis** blood factors are not entirely clear. Perhaps the main reason for this is that up to now it has been impossible to prepare reagents of satisfactory avidity and titer. The reagents in general have the properties, as mentioned above, of cold agglutinins of low titer, and generally have associated with them non-specific cold agglutinins that are difficult to remove even by absorption. Nevertheless, certain interesting facts have emerged.

In 1948, Grubb[60] reported that practically all persons who have the $\mathbf{Le^a}$ factor on their red cells are non-secretors of the A-B-O blood group substances, while those lacking the $\mathbf{Le^a}$ factor almost always are secretors of the A-B-O blood substances. Tests on the saliva of $\mathbf{Le^a}$-positive individuals, using the inhibition technique, have shown that the saliva can neutralize anti-$\mathbf{Le^a}$ serums, suggesting that the Lewis substance takes the place of the A-B-O substance in appropriate instances. In this connection it is interesting to note that chemical studies by Morgan et al.[61] prove that the Lewis substance is a mucopolysaccharide chemically indistinguishable by present methods from the A-B-O substances.

Family studies on heredity of the $\mathbf{Le^a}$ factor show that the transmission of $\mathbf{Le^a}$ parallels that of the secretor/non-secretor trait. Thus two Le(a+) parents, who would both be non-secretors of the A-B-O blood group substances, will have children all of whom are Le(a+) and non-secretors of the A-B-O substances. This expectation has been fulfilled by family studies carried out by Race et al.,[62] and by Andresen et al.,[63] and others, who in 17 such families found 44 children all of whom were Le(a+). On the other hand, when both parents are Le(a−), children who are Le(a+) as well as Le(a−) occur in a ratio consistent with the hypothesis that factor $\mathbf{Le^a}$, in contrast to other blood factors, behaves in inheritance like a Mendelian recessive.

Studies on the heredity of the $\mathbf{Le}^a$ factor are complicated by the variable reactivity of the red cells at different ages. Andresen early observed that as many as 80 per cent of infants less than 3 months old had Le(a+) red cells. Among older babies the percentage of positive reactions was lower, but not until the end of the first year of life did the frequency drop to the adult 20 per cent level. On the other hand, Unger found that cord blood specimens are all Le(a−), and according to Rosenfield and Ohno, the baby's cells become agglutinable only after the first few weeks of life. Grubb and Morgan observed, moreover, that the more Le(a+) cells are washed with saline solution, the less agglutinable by anti-$\mathbf{Le}^a$ serum do they become. However, Race and Sanger were unable to wash the Le^a antigen off the red cells and they suggest that such observations may depend on the antiserum used.

Sneath and Sneath[64] reported that when Le(a+) persons were transfused with Le(a−) blood, the donor's red cells could not be demonstrated in the recipient's circulation by differential agglutination tests employing anti-$\mathbf{Le}^a$ serum, although when antiserums which detected other agglutinogens were used the donor's cells could readily be demonstrated. This discrepancy in the results of the differential agglutination tests could be explained by postulating that the donor's Le(a−) cells adsorbed the Le^a substance present in the serum of the recipient and thus became Le(a+). This suggests that in contrast to other blood group substances, the Lewis substance is primarily produced in the tissues and body fluids, and secondarily absorbed onto the red cells. Additional evidence supporting this idea was obtained by studies on a rare instance of mosaicism in a pair of fraternal twins.[65] The blood vessels of these twins contained a mixture of two kinds of red blood cells having the genetic blood groups of the two twins respectively. The red cells of the twins reacted in differential agglutination tests like blood obtained from individuals who have had blood transfusions, namely, like a mixture of two types of blood. However, the twins failed to exhibit mosaicism of their Lewis blood types; one twin was a non-secretor of A-B-O group substance and had red cells all Le(a+), while the other, a secretor, had cells all Le(a−), even though their red cells exhibited mosaicism of other blood group systems.

When red cells are tested with both sera anti-$\mathbf{Le}^a$ and anti-$\mathbf{Le}^b$, conflicting results are obtained. One cause of difficulty is that anti-$\mathbf{Le}^b$

serums give many more positive reactions with cells of group O and group A_2 than with cells of group A_1 and B. When tests are confined to blood cells of group O and A_2, **Le**^{a} and **Le**^{b} prove to be almost but not completely antithetical. For example, among 238 group O individuals, Andresen found 46 Le(a+ b−), 178 Le(a− b+), 14 Le(a− b−) and no Le(a+ b+). Other investigators have confirmed these findings, and although a number of genetic theories have been proposed none satisfactorily explains these findings.

In conclusion, the following hypothesis is offered in an attempt to account for some of the puzzling observations regarding the relationship between the Lewis blood types and the A-B-O blood group substances and secretor types. It is suggested, firstly, that the substance Le^a is identical with the substance T found in non-secretor saliva by Uyeyama, and that substance Le^b is the same as or related to the substance H of secretor saliva,* while both substances are believed to be modifications of a single basic mucopolysaccharide. If this is so, then individuals who are genetically non-secretors (genotype *sese*) would be expected to have the substance T (or Le^a) in their secretions and body fluids while secretors (genotype *SeSe* and *Sese*) will have substance H (or Le^b), although heterozygotes may have small amounts of Le^a (or T) in addition. The group substances of secretors, in addition to factor H will have factors **A** or **B**, or both, or neither, depending on the A-B-O blood group. The genetically determined group specific substance (mucopolysaccharide) present in the body fluid is absorbed onto the red cells, which thus may acquire the **Le**^{a} (or **T**), or **Le**^{b} (or **H**) factor according to the secretor type, in addition to the group properties originally present. According to this hypothesis, one might anticipate that the red cells of group O individuals who are secretors should react more strongly with anti-**H** serums than the red cells of non-secretors. Indeed, some preliminary tests using extracts of Ulex europaeus as anti-**H** reagents appear to support this expectation.

Theoretico-Statistical Considerations. As has already been pointed out, the presence of **Le**^{a} factor on the red cells is demonstrable only in homozygous individuals, genotype Le^aLe^a, bearing in mind

* According to W. T. J. Morgan (personal communication, however, some preparations of H substance have Le^b activity, while other preparations do not.

TABLE 9

THE LEWIS TYPES*

Red Cell Phenotypes		Corresponding Genotypes		Four Salivary Types	
Designation	Observed frequencies (per cent)	Designations†	Theoretical frequencies	Designation	Estimated frequencies (per cent)
Le(a −b−)	6.01	*lele*	c^2	L_o	0.4
		Le^a*le*	$2ac$	L_a	28.0
Le(a +b−)	22.38	*Le*^{a}*Le*^{a}	a^2		
Le(a −b+)	71.61	*Le*^{b}*le*	$2bc$	L_b	27.4
		Le^b*Le*^{b}	b^2		
		Le^a*Le*^{b}	$2ab$	L_{ab}	44.2

*Modified after Wiener [*Lab. Digest* (St. Louis) Vol. 18, No. 5, Nov., 1954].

† Gene *Le*^{b} is believed to be identical with gene *Se*; similarly, genes *Le*^{a} and *le* may both correspond to gene *se*.

that gene *Le*^{a} and gene *se* for the ABH non-secretor trait may be identical.[66] This may represent a gene dosage effect, since all the available anti-**Le**^{a} serums have been of very low titer and avidity. (Similarly, weak anti-**N** serums fail to agglutinate type MN cells while clumping type N cells distinctly.) In secretors of the ABH substance, the group substances have proved to be present in saliva, gastric juice and other secretions in considerably higher concentrations than on the red cells. If the same is true of the Le^a substance, then positive reactions by the inhibition test on saliva would be expected in heterozygous as well as homozygous **Le**^{a}-positive individuals. Thus, bearing in mind that the **Le**^{b} and **H** factors are closely related, and that genes *Le*^{b} and *Se* may be identical, the distribution of the Lewis types as determined by tests on saliva would be expected to be different from that obtained by testing red cells.

Let L_0, L_a, L_b, and L_{ab} represent the four Lewis types as determined by testing saliva, where L_0 represents absence of both substances, L_a represents presence of Le^a and absence of Le^b, etc. Also, let a, b, and c represent the frequencies of genes *Le*^{a}, *Le*^{b}, and *le*, respectively, where *Le*^{a} determines the presence of Le^a substance, *Le*^{b} determines the presence of substance Le^b, and *le* determines neither. Then table 9 can be drawn up.[67]

Table 9 also gives the distribution of the three Lewis red cell phenotypes in England, as reported by Race and Sanger[1e] in their book. From the frequencies of these three phenotypes the frequencies of the three Lewis genes can be estimated as follows:

$a = \sqrt{\text{Le(a+ b−)}} = \sqrt{0.2238} = 47.3$ per cent

$c = \sqrt{\text{Le(a− b−)} + \text{Le(a+ b−)}} - \sqrt{\text{Le(a+ b−)}} = 6.0$ per cent,

And, $b = 1 - a - c = 46.7$ per cent

From these estimated gene frequencies the expected distribution of the four salivary Lewis types can be calculated as follows:

$L_0 = c^2 = (.06)^2 = 0.36$ per cent

$L_a = a^2 + 2ac = (0.473)^2 + 2(0.473)(.06) = 28.05$ per cent

$L_b = b^2 + 2bc = (0.467)^2 + 2(0.467)(.06) = 27.4$ per cent

$L_{ab} = 2ab = 2(0.473)(0.467) = 44.2$ per cent

Unfortunately, very little data are available with which to test these calculations. Grubb and Morgan[66] have reported observations on the correlation between the Le^a phenotype of the red cells and the ABH secretory type in a series of 122 individuals as follows:

Erythrocytes Le(a+), ABH non-secretor....................	57
Erythrocytes Le(a−), ABH secretor......................	163
Erythrocytes Le(a+), ABH secretor......................	0
Erythrocytes Le(a−), ABH non-secretor..................	2

If we assume that the two non-secretors who were of red cell phenotype Le(a−) had no Lewis substance in their saliva, this would make the frequency of the L_0 type only 2 among 222 individuals or 0.9 per cent which does not deviate significantly from the expected frequency of 0.4 per cent.

Quite different results have been reported by Ceppellini, who tested the saliva of 518 individuals for the ABH and Le^a substances. He designated ABH secretors as S and nonsecretors as ns; similarly Le^a secretors were designated as L and non-secretors as nl. Ceppellini[68] thus identified 4 salivary types with the following distribution among 518 individuals:

ns nl..................	11 or 2.13 per cent
ns L..................	83 or 16.02 per cent
S nl..................	50 or 9.25 per cent
S L..................	374 or 72.25 per cent

When these findings were arranged in a 2 × 2 contingency table, Ceppellini found $\chi^2 = 0.036$, which in his opinion strongly indicated that the secretions of the Le^a and ABH substances were genetically independent of one another. If, as seems likely, the four types ns nl, ns L, S nl, and S L, correspond to the salivary types L_0, L_a, L_b, and L_{ab}, respectively, this would mean that the heredity of Le^a and Le^b depends on independent pairs of genes rather than corresponding allelic genes. Thus, the conflict between Ceppellini's findings and views and the hypothesis favored by the authors is reminiscent of the polemic between Hirszfeld and Bernstein regarding the heredity of the four A-B-O groups. A serious objection to Ceppellini's hypothesis* is that it does not provide a satisfactory explanation for the reciprocal relationship between the Le^a blood type and the ABH secretor type.* On the other hand, since the frequency 72.25 per cent which he observed for type S L or L_{ab} significantly exceeds 50 per cent, this would argue against the concept of Le^a and Le^b as allelic genes. However, Ceppellini's observations conflict with those of Grubb and Morgan, and the explanation possibly lies in the difficulty of testing for Le^a substance is saliva. As already pointed out, the anti-$\mathbf{Le}^a$ reagents available are of low titer and avidity, so that inhibition tests with such reagents do not give clear cut results. In fact, examination of sample protocols given by Ceppellini shows no sharp division between types L and nl so that not infrequently the diagnosis seems arbitrary. Thus, the introduction of a slight bias when reading the reactions would strongly influence the results obtained. Obviously, until better reagents become available, judgment must be suspended.

A third hypothesis has been suggested by Grubb and Morgan. They suggest that there are three pairs of allelomorphic genes, *Se* and *se*, Le^a and Le^α, Le^b and Le^β, occupying contiguous loci. This hypothesis differs from the one proposed here in the same way as Fisher's theory of triply linked genes differs from Wiener's theory of multiple alleles for the Rh-Hr blood types. In the absence of evidence of crossing over, the theory of separate gene loci degenerates into a theory of complete

* To account for the existence of only 3 Lewis blood types, Ceppellini invokes the concept of gene interaction. The gene combination *LS* is postulated to result in type Le(a− b+), *Ls* in Le(a+ b−), while *LS* and *ls* both give Le (a− b−). Ceppellini's saliva results then give the following expected distribution of the Lewis blood types: 16.02% Le(a+ b−), 72.25% Le(a− b+), and 11.38% Le(a− b−).

linkage, and since completely linked genes operate as a unit, the difference from the theory of multiple alleles becomes solely a matter of semantics.

A-B-O Blood Groups in Apes and Monkeys

In 1925, Landsteiner and Miller[69] showed that blood of chimpanzees, orangutans and gibbons have isoagglutinogens and isoagglutinins indistinguishable from those which characterize the human blood groups. To date more than 100 chimpanzees have been tested, of whom approximately 90 per cent proved to belong to group A and the remainder to group O. Group A chimpanzee blood resembles human blood of subgroup A_1 more closely than subgroup A_2.[70] Many fewer of the other anthropoid apes have been tested; among 18 orangutans, 7 proved to be group A, 8 group B, and 3 group AB. Among 7 gibbons, Landsteiner found 1 group A, 5 group B, and 1 group AB.

Wiener tested the blood of two dead gorillas and found anti-**A** in the serums, but tests on the cells failed to reveal the presence of the expected agglutinogen B, in conflict with Landsteiner's law. Examination of the salivary glands, however, showed them to contain blood group substance B. In studies on various species of monkeys, Wiener, Candela and Goss[71, 72] found that their red cells were not agglutinated by anti-**A** or anti-**B** serums, but the serum of individual monkeys differed in that some contained anti-**A** alone, some anti-**B** alone, others showed both anti-**A** and anti-**B**, while the remainder showed neither. Examination of the saliva and salivary glands explained these observations since the expected blood group substances were found to be invariably present. Thus, in gorillas and monkeys Landsteiner's law does hold, but the reciprocal relation is between the isoagglutinins in the serum and the blood group substances in glandular tissue and secretions rather than on the red cells themselves.

Anomalies of the A-B-O Blood Groups

Recently, Wiener and Gordon[73] encountered two individuals having anti-**B** and not anti-**A** agglutinins in their serums, but whose red cells failed to react with anti-**A** serum. Examination of the saliva showed the presence of the expected blood group substance A. Thus, these two rare individuals gave blood group reactions resembling those of lower monkeys, and for this reason this anomalous blood type has been designated A_m. Additional examples of this blood type have been observed by Weiner et al.,[74] who postulated the existence of a pair of allelic genes *Y* and *y* independent of the A-B-O genes, where *y* is extremely rare. The extremely rare homozygotes of genotype *yy* are postulated to be capable of producing blood group substances in the secretions but not on the red blood cells. The counterpart of A_m, namely, B_m, has recently been encountered in a father and 2 of his 4 children by Yokoyama et al.[75]

In 1951, Bhende et al.[76] described a very rare type of blood which has been called "Bombay". The red cells of individuals of this type were not clumped by anti-**A**, anti-**B**, or anti-**H** and the serum contained anti-**H** as well as anti-**A** and anti-**B**. The saliva of these persons was devoid of A-B-O group specific substance. Levine et al.[77] have reported a family in which three members had the "Bombay" blood type. From the blood groups of the parents and children they were led to postulate the existence of a rare modifying gene which in homozygous individuals, genotype *xx*, prevents the blood group genes from producing A-B-O specific substances both on the red cells and in the saliva. The red cells of all these individuals are Le(a+), which is further evidence of the reciprocal relationship between **Le** and **H**. Levine has assigned to blood of the Bombay type the symbol, O_h, to indicate that the red cells react like group O cells, but lack the blood factor **H**.

The recent developments demonstrate that the heredity of the A-B-O blood groups is quite complex, and the suggestion made by Haldane that the blood group substances are the direct products of the blood group genes does not seem tenable in the light of these observations. Apparently, all or almost all humans are capable of producing a basic blood group mucopolysaccharide, and the capacity to produce this substance presumably depends upon the action of two or more pairs of intermediate genes. The blood group genes I^A, I^B and I^O apparently convey the capacity to produce certain enzymes which act on the basic mucopolysaccharide and give it its blood group specificity. Still another pair of genes *Se* and *se* determine the capacity to produce the polysaccharide in glandular tissue as well as in the erythrocytes and body cells, and confer $\mathbf{Le^b}$ (or **H**) and $\mathbf{Le^a}$ (or **T**) specificity on these substances. Individuals of the very rare genotype *yy* apparently can produce group specific substances in the secretions but not on the red blood cells, while the mucopolysaccharides of individuals of the very rare genotype xx ("Bombay" type) completely lack ABH group specificity both in secretions and on red cells. The presence of genotype xx, while it prevents the expression of the blood group, does not interfere with the hereditary transmission of the A-B-O genes, so that the blood group will show up as expected in other individuals of the family who are not homozygous for gene x.

CHAPTER III
The M-N-S System

M-N Types

The three M-N types provide one of the simplest examples of Mendelian heredity in man.

Serology. To determine the M-N types of an individual, anti-**M** and anti-**N** reagents from rabbit-immune antiserums are still used today, as in the original investigations of Landsteiner and Levine.[78, 80] Occasionally, human serums have been encountered containing anti-**M** and anti-**N** agglutinins, but these are so rare that their use is not practicable for routine blood typing. Similarly, adequate quantities of seeds yielding anti-**N** lectins have not been generally available for this purpose so far, while anti-**M** has not been found in seeds.

The anti-**M** serums produced by immunizing rabbits with human type M blood contain in addition to anti-**M** also anti-human species-specific agglutinins. The serum must therefore be processed by absorbing it with type N cells, which removes the species-specific antibodies but leaves behind the anti-**M** agglutinin. Anti-**N** serum is processed similarly by absorbing it with type M cells. While there is little difficulty in preparing potent anti-**M** reagents in this way, it has not been possible to obtain high titer anti-**N** reagents. The reason is that the type M blood cells used to absorb the species antibody can also absorb some of the desired anti-**N** agglutinin; in fact, if an excess of cells is used the reagent may be made devoid of activity or weakened so that only type N, but not type MN cells are agglutinated. (Also anti-**N** agglutinins in human serums can be absorbed by type M cells.*) The practical importance of this phenomenon is that there is danger of obtaining false negative reactions if the anti-**N** reagents have been overabsorbed. On the other hand, if the serum has been underabsorbed in an attempt to avoid the removal of the type specific agglutinin, reagents may result which give false positive reactions. This difficulty has been responsible for serious errors in the application of the M-N types to anthropology and legal medicine, when the

* This is another of the numerous examples of the cross reactivity of blood group antibodies, as discussed on page 11.

tests were undertaken by workers who did not first qualify themselves by becoming experienced in serologic techniques.

Anti-**M** and anti-**N** reagents yield only three pairs of reactions with human red cells, namely, M+ N−, M+ N+, and M− N+, corresponding to the three types designated M, MN, and N, respectively. Bloods lacking both agglutinogens have not been encountered in tests on many millions of blood specimens.

Heredity. According to the theory of Landsteiner and Levine,[80] the agglutinogens M and N are transmitted by two corresponding allelic genes L^M and L^N. This theory allows for only three possible genotypes corresponding to the three known phenotypes, and at the same time precludes the existence of type M− N−. There are six matings theoretically possible and the distribution of the types to be expected among the children of each mating is shown in table 10. While most studies conform with the predictions, some workers have reported an excess of type MN children in those families where both parents are type MN, and this supposed excess of type MN children has been cited as an example of selection in favor of the heterozygote. In table 11, is shown the summary of the investigations carried out by Wiener and his collaborators during the years 1929–52 on 1,580 families with 3,379 children. Among the 800 children from the mating MN × MN, 405 belonged to group MN, a finding consistent with simple random redistribution of the genes L^M and L^N, without the need to invoke other mechanisms. The excess of type MN children reported by other workers was therefore probably due to the use of incompletely absorbed antiserums, as already explained.

From the point of view of medicolegal application, the heredity

TABLE 10

HEREDITY OF THE M-N TYPES

Matings		Children (per cent)		
Phenotypes	Genotypes	M (L^ML^M)	MN (L^ML^N)	N (L^NL^N)
M × M	$L^ML^M \times L^ML^M$	100	0	0
M × N	$L^ML^M \times L^NL^N$	0	100	0
N × N	$L^NL^N \times L^NL^N$	0	0	100
M × MN	$L^ML^M \times L^ML^N$	50	50	0
N × MN	$L^NL^N \times L^ML^N$	0	50	50
MN × MN	$L^ML^N \times L^ML^N$	25	50	25

TABLE 11

HEREDITY STUDIES ON THE M-N TYPES (1929–1952)*

(After Wiener, A. S., DiDiego, N. and Sokol, S.: *Acta Genet. Med. et Gemellol.*, **2:** 391, 1953)

Parental Combination	Number of Families		Children Belonging to Type			Total
			M	N	MN	
M × M	153	Number	326	0	(1)†	327
		Per cent	99.7	0	0.3	
M × N	179	Number	(1)†	0	376	377
		Per cent	0.3	0	99.7	
N × N	57	Number	0	106	0	106
		Per cent	0	100.0	0	
MN × M	463	Number	499	(1)†	473	973
		Per cent	51.3 ± 1.1‡	0.1	48.6 ± 1.1	
MN × N	351	Number	(3)†	382	411	796
		Per cent	0.4	47.9 ± 1.2	51.7 ± 1.1	
MN × MN	377	Number	199	196	405	800
		Per cent	24.7 ± 0.8	24.5 ± 1.0	50.8 ± 1.1	
Totals	1,580		1,028	685	1,666	3,379

* Since that time many hundred additional families have been studied.

† These apparent exceptions to the laws of inheritance are probably due to illegitimacy.

‡ Figure following ± sign represents probable error.

of the M-N types may be summarized by the following two laws: 1) The agglutinogen M or N cannot appear in the blood of a child unless it is present in the blood of one or both parents. 2) A type M parent cannot have a type N child, and a type N parent cannot have a type M child. The few exceptions to these rules in table 10 may reasonably be attributed to illegitimacy, and in support of this interpretation it may be pointed out that while there are families in which the blood types of the supposed fathers conflict with the second law, there are none in which this is true of the mother.

Gene Frequency Analysis. The frequency of the allelic genes L^M and L^N in the general population can be readily estimated from the distribution of the three phenotypes.[81] Since there is only one phenotype corresponding to each genotype, the gene frequencies can be determined by direct count. If m represents the frequency of gene L^M, and n represents the frequency of gene L^N,

$$\text{Then,} \quad m = \overline{\text{M}} + \frac{\overline{\text{MN}}}{2} \tag{21}$$

$$\text{And,} \quad n = \overline{\text{N}} + \frac{\overline{\text{MN}}}{2} \tag{22}$$

If V represents the size of the sample, then $2V$ is the number of genes counted.

$$\text{Therefore,} \quad \sigma_m = \sigma_n = \sqrt{\frac{mn}{2V}} \tag{23}$$

From the gene frequencies, the distribution of the three M-N types can be recalculated as follows:

$$\overline{\text{M}}_0 = m^2 \tag{24}$$

$$\overline{\text{N}}_0 = n^2 \tag{25}$$

$$\overline{\text{MN}}_0 = 2mn \tag{26}$$

As a rule, these recalculated frequencies will not be identical with the observed frequencies. Assuming random intermarriage, which theoretically should bring the distribution of the M-N types to equilibrium in a single generation, the genetic theory can be tested by calculating the chi square value as follows.[81]

$$\chi^2 = \frac{(a - a_0)^2}{a_0} + \frac{(b - b_0)^2}{b_0} + \frac{(c - c_0)^2}{c_0}$$

Here, a, b, and c represent the observed frequencies of the types M, MN, and N, respectively, in the population, and a_0, b_0, and c_0 the theoretically expected values according to formulae (24) to (26). This then reduces to the following formula:

$$\chi^2 = \frac{(b^2 - 4ac)^2 N}{(2a + b)^2(b + 2c)^2} \tag{27}$$

where N is the number of individuals examined. This test has been applied in a large number of population studies, and the results lend further support to the genetic theory.

The frequencies of the genes L^M and L^N can also be calculated by the square root method as follows:

$$m' = \sqrt{\overline{\text{M}}} \tag{28}$$

$$n' = \sqrt{\overline{\text{N}}} \tag{29}$$

These formulae for the gene frequencies, though consistent, are less efficient than formulae (21) and(22), the latter being identical with the maximum likehood estimates.

The formulae (27) and (28) are the basis for a different test of the genetic theory. Since theoretically, $m' + n' = 1$ then $\sqrt{\overline{\text{M}}} + \sqrt{\overline{\text{N}}}$ should equal unity or 100 percent. In general, the sum $\sqrt{\overline{\text{M}}} + \sqrt{\overline{\text{N}}}$ will not be exactly equal to unity, and the size of the deviation can be used as a test of the genetic theory. If the deviation:

$$D = 1 - \left(\sqrt{\overline{\text{M}}} + \sqrt{\overline{\text{N}}}\right) \tag{30}$$

Then, as Wiener[81] has shown,

$$\sigma_D = \frac{1}{2\sqrt{V}} \tag{31}$$

where V represents the number of persons tested. Here again the observations confirm the genetic theory.

If we represent the frequency of type MN by H,

$$\text{Then,} \quad \frac{dH}{dm} = \frac{d}{dm}\, m(1 - m) = 1 - 2m$$

To find the maximum value of H, set $\frac{dH}{dm} = 0$, so that $m = \frac{1}{2}$.

Therefore, the maximum frequency that type MN can have under this genetic theory is 50 per cent.

This important result applies to any heterozygote in a population at equilibrium, so that, for example, the frequency of group AB cannot exceed 50 per cent. As will be pointed out later, this law was applied by Wiener to elucidate the heredity of the Rh-Hr blood types (cf. page 63).

Mother-Child Combinations. As already pointed out, the study of mother-child combinations supports the theory because of the absence of the combinations M-mother/N-child or N-mother/M-child. Further evidence supporting the theory has been obtained

TABLE 12

FREQUENCIES OF MOTHER-CHILD COMBINATIONS

Mothers of Type	Children of Type		
	M	MN	N
M	m^3	m^2n	0
MN	m^2n	mn	mn^2
N	0	mn^2	n^3

by gene frequency analysis, by applying the formulae of table 12 to test the results of studies on the M-N types in mothers and children.

Agglutinogen N_2. As in the A-B-O system, a number of rare anomalies of the M-N types have been encountered. The most important is the so-called agglutinogen N_2, first described by Crome,[82] and which, according to Friedenreich,[83] is due to a special corresponding rare allelic gene L^{N_2}. This type of blood is characterized by its very weak reactivity with anti-**N** serums, so that with many anti-**N** reagents clumping may fail to occur or may be so weak as to be overlooked. This may lead to serious errors, as in medicolegal cases. For example, if the putative father belongs to type MN_2 (genotype $L^ML^{N_2}$) and the child to type N (genotype $L^NL^{N_2}$), the former might erroneously be reported as type M, with resulting false exclusion of paternity.

Agglutinogens* M *and* N *in Monkeys and Apes. In tests on human blood no differences are demonstrable among different anti-**M** reagents or different anti-**N** reagents. For example, all anti-**M** serums, regardless of source and method of preparation, regularly agglutinate red cells of type M and of type MN but not of type N. However, when the cross reactions of these reagents are studied, using red cells of anthropoid apes and monkeys, it is found that not all anti-**M** serums give the same reactions, nor do all anti-**N** serums.[84, 85]

In table 13, for example, are shown the differences in behavior of six selected anti-**M** serums prepared from immune rabbit serums, in tests on red cells from anthropoid apes and monkeys. In the table, the antiserums have been arranged so that those with more cross reactions precede those with fewer cross reactions, and the blood specimens have been arranged according to the position in the zoölogical scale of the primates from which they were obtained. Although the anti-**M** serums with the highest titers gave the most cross reactions, the differences were not merely quantitative as was shown by absorption experiments. For example, reagent M1 had a titer of 64 units for human red cells and 16 units for rhesus red cells. Absorption by human cells completely removed all the antibodies while absorption by rhesus cells produced a reagent with a titer of 16 units for human M cells but not reacting with rhesus cells. On the other hand, reagent M5 which likewise had a titer of 64 units for human M cells and about 24

TABLE 13

M FACTORS IN BLOOD OF APES AND MONKEYS

(After Wiener, A. S.: *Blood Groups and Transfusion*, 3rd ed., 1943)

Source of Blood Specimens	Anti-**M** Testing Fluids					
	M5	M1	M21	M35	M2	M82
Human M	+++	+++	++±	++±	++±	++±
Human N	0	0	0	0	0	0
Chimpanzee	+++	+++	+++	++	+++	±
Old World Monkeys (Cercopithecidae)						
Sphinx Baboon	++|	++	++±	0		0
Drill Baboon	+++	++|	++±	(±)	(+)	(+±)
Chacna Baboon	+++	+++	++±	0	tr.	0
M. rhesus	+++	+++	++±	(+±)	±	0
Java Macaque	+++	+++	+±	0	0	0
Sooty Mangabey	+++	+++	++±	tr.	±	±
Green Monkey	+++	+++	0	0	0	0
New World Monkeys (Platyrrhina)						
White Spider Monkey	++±	0	0	0	0	0
Black Spider Monkey	±	0	0	tr.		0
Woolly Monkey	0	0	(±)	0	0	0
Brown Ringtail (Capuchin Monkey)	0	0	0		0	0
Moss Monkey	0	0	f. tr.	±		0
Lemur	0	0	0	0	0	0
Average titer of testing fluids	64	64	32	24	16	16

Of several species two individuals were tested, of brown ringtails 4, of Macacus rhesus 45.

Reactions placed in parenthesis were found not to be removed by absorbing the serum with type M blood cells.

units for rhesus red cells was completely absorbed by either human M cells or rhesus red cells.

Besides providing a striking example of biochemical evolution, table 13 is another example of the mosaic structure of red cell agglutinogens. To explain the reactions, it is necessary to recognize that not all anti-**M** agglutinins are identical in specificity and therefore, by definition, there are more than one kind of **M** factor. If we designate as anti-$\mathbf{M_i}$ the agglutinin which reacts with human red cells, but not with blood from chimpanzees or monkeys, and as anti-$\mathbf{M_{ii}}$ the agglutinin which detects a factor common to human M blood and chimpanzee blood but absent from the blood of lower primates, etc., it follows that the human M agglutinogen has all of the blood factors $\mathbf{M_i M_{ii} M_{iii}}$ . . . etc.; chimpanzee blood, factors $\mathbf{M_{ii} M_{iii}}$ . . ., etc. One could anticipate that the injection of rhesus monkey blood into rabbits would yield a reagent specific for human M cells and not reacting with type N cells. Indeed, Landsteiner and Wiener[84] found it possible to prepare satisfactory anti-**M** reagents in this way, and it is of historical interest that these experiments led to the discovery of the Rh (rhesus) factor.

By similar experiments it has been possible to demonstrate that the agglutinogen N has a mosaic structure.[84] Due to difficulty in obtaining potent reagents, as already explained, only two such factors have been demonstrated so far, namely, $\mathbf{N_i}$, peculiar to human N cells alone, and $\mathbf{N_{ii}}$, a factor shared by human N cells and chimpanzee red cells.

Obviously, when a human being inherits an M or N agglutinogen, this carries with it all of the serological factors that characterizes the agglutinogen as shown below:

Gene	Corresponding agglutinogen	Blood factors
L^M	M	$\mathbf{M_i}$, $\mathbf{M_{ii}}$, $\mathbf{M_{iii}}$, $\mathbf{M_{iv}}$. . .
L^N	N	$\mathbf{N_i}$, $\mathbf{N_{ii}}$. . .

Factors *S*, *s* and *U*

Factor S. In 1947, Walsh and Montgomery[86] studied the serum of a patient known to be sensitized to the $\mathbf{Rh_0}$ factor, and found the serum to contain an additional blood group antibody which gave reactions different from any previously described. A sample of this serum was studied by Sanger and Race[87] who were able to show that the blood factor detected by it was related to the agglutinogens M

TABLE 14

DISTRIBUTION OF THE M-N-S TYPES AMONG 394 BLOOD DONORS

Phenotype		Number	Per Cent
M	S+	88	22.3
	S−	37	9.4
N	S+	27	7.0
	S−	49	12.4
MN	S+	106	26.9
	S−	87	22.0
Totals	S+	221	56.2
	S−	173	43.8

and N. This was done by making use of the 2 × 2 contingency table as already explained on page 3.

When persons of type M, MN and N were subdivided according to their reactions with the antiserum of Walsh and Montgomery, named anti-**S** by Sanger and Race, 6 types of blood could be distinguished. For example, in 394 tests done on professional donors in New York City, the distribution shown in table 14 was found.[88]

On inspection it can be seen that the incidence of the **S** factor is not the same in the three M-N types, since more than two-thirds of type M individuals are **S** positive, while only about one third of type N individuals are **S** positive. That this difference is not merely accidental but statistically significant can be shown by a 2 × 2 contingency table as follows:

	M + MN	N	
S+	194	27	221
S−	124	49	173
Totals.............	318	76	394

Applying the formula on page 3, $\chi^2 = 16$ for one degree of freedom, and the likelihood of this occurring by chance alone is much less than 0.001.

To account for the findings, Sanger and Race mention two genetic hypotheses. According to one, the six possible M-N-S groups are at-

TABLE 15

THE M-N-S PHENOTYPES AND GENOTYPES

Phenotypes	Reactions with Serums			Corresponding Genotypes	
	Anti-**M**	Anti-**N**	Anti-**S**	Linked genes	Multiple alleles
M.S	+	−	+	*MS/MS* & *MS/Ms*	L^SL^S & L^SL
M.s	+	−	−	*Ms/Ms*	LL
N.S	−	+	+	*NS/NS* & *NS/Ns*	l^Sl^S & l^Sl
N.s	−	+	−	*Ns/Ns*	ll
MN.S	+	+	+	*MS/NS*, *MS/Ns* *Ms/NS*	L^Sl^S, L^Sl & Ll^S
MN.s	+	+	−	*Ms/Ns*	Ll

tributed to 4 allelic genes *MS*, *Ms*, *NS*, and *Ns*, where *S* is considered to be a mutation in the *M* and *N* genes which renders the red cells agglutinable by anti-**S** serum. According to the second hypothesis, *S* is considered to be a separate gene, with an allele *s* and closely or possible completely linked with the M-N genes. The six possible genotypes, their reactions and corresponding genotypes according to each of these hypotheses are shown in table 15. Though Race and Sanger prefer the linked gene theory, the present authors regard the concept of multiple alleles to be more consistent with the known genetic and serological data.[89]

When factor **S** is considered alone, two types are distinguishable, namely, S+ and S−, the former having the frequency of 56.1 per cent and the latter 43.9 per cent in tests done on 394 blood donors in New York City. Assuming heredity by a pair of genes *S* and *s*, so that corresponding to phenotype S there are two genotypes *SS* and *Ss*, and corresponding to phenotype s there is only one genotype *ss*, the frequencies of genes *S* and *s* can be estimated as follows.

$$s = \sqrt{0.439} = 0.663 \text{ or } 66.3 \text{ per cent}$$

so that $S = 0.337$ or 33.7 per cent.

Therefore, the estimated frequencies of the three genotypes are $SS = 0.114$, $Ss = 0.447$, and $ss = 0.439$.

At this stage of knowledge, it could be surmised that there might exist a factor, **s**, corresponding to the gene *s*, and demonstrable in the blood cells of individuals of genotypes *Ss* and *ss*. In fact, Levine et al.[90] did find a serum in the mother of an erythroblastic infant which gave reactions corresponding to those expected for factor **s**. Subsequently, additional examples of this antiserum were encountered by other workers. However, while a fair number of satisfactory anti-**S** serums have been found to date, serum of the specificity of anti-**s** is in very short supply and more or less of a serological curiosity. Therefore, the number of studies carried out on factor **s** have been quite small.

Nomenclature. The vast majority of studies on the M-N system have been confined to the blood factors **M** and **N** alone, since these have been known for the longest time and the reagents for their recognition can be procured at will by immunizing rabbits. As mentioned previously, three types are distinguished with the aid of these antiserums. When tests for factor **S** are also carried out, the number of phenotypes in the system is increased to six. When, in addition, anti-**s** serum is used, nine phenotypes can be recognized. Since an increasing number of studies will be made in the future, using the various serums at different stages, it is desirable to have an unambiguous nomenclature for each of the three stages, which at the same time will indicate clearly not only the phenotypes but also the theoretically possible genotypes. The nomenclature used by Wiener[91] is shown in table 16.

TABLE 16
NOMENCLATURE OF THE M-N-S-s TYPES
(After Wiener)

Three M-N Phenotypes			Six M-N-S Phenotypes		Nine M-N-S-s Phenotypes		Corresponding Genotypes
Designation	Reaction with serum		Designation	Reaction with serum	Designation	Reaction with serum	
	Anti-**M**	Anti-**N**		Anti-**S**		Anti-**s**	
M	+	−	M.S	+	M.SS	−	L^SL^S
					M.Ss	+	L^SL
			M.s	−	M.ss	+	LL
N	−	+	N.S	+	N.SS	−	l^Sl^S
					N.Ss	+	l^Sl
			N.s	−	N.ss	+	ll
MN	+	+	MN.S	+	MN.SS	−	L^Sl^S
					MN.Ss	+	L^Sl and Ll^S
			MN.s	−	MN.ss	+	Ll

Gene Frequency Analysis.[91, 92] If the distribution of the M-N-S phenotypes in the general population is known, the hypothetical gene frequencies can be readily calculated.

Stage 1. At this stage only tests for factors **M** and **N** have been done, and three phenotypes distinguished. In this case, the gene frequencies, m and n can be readily calculated, as previously explained, by direct count.

Stage 2. Here tests have been made for factor **S** as well as for **M** and **N**. The following relations hold by identity:

$$L^S + L = m \tag{32}$$

$$l^S + l = n \tag{33}$$

where m and n are the frequencies of gene L^M and gene L^N respectively [cf. formulae (21) and (22)].

Moreover, $L^S + l^S = S$ (34)

and $L + l = s$ (35)

Since, corresponding to the phenotype N.s there is only the single genotype ll, then, assuming panmixia, it follows that

$$l = \sqrt{\overline{\text{N.s}}} \tag{36}$$

where $\overline{\text{N.s}}$ represents the frequency of type N.s in the general population.

Similarly, $L = \sqrt{\overline{\text{M.s}}}$ (37)

These preliminary estimates must now be adjusted to make them satisfy the identity $L + l = s$. This can be done by solving the simultaneous equations,

$$L - l = \sqrt{\overline{\text{M.s}}} - \sqrt{\overline{\text{N.s}}} \tag{38}$$

$$\text{and } L + l = s \tag{39}$$

where $s = \sqrt{\overline{\text{type s}}} = \sqrt{\overline{\text{M.s}} + \overline{\text{N.s}} + \overline{\text{MN.s}}}$ (40)

From these values of L and l, the values of L^S and l^S can be obtained with the aid of identities (30) and (31).

Stage 3. Here tests have been carried out with all four antiserums, so that 9 phenotypes are distinguishable (cf. table 16). The calculations are the same as at stage 2, except that now the frequency of "gene" *s* can be determined by direct count as follows.

$$
\begin{aligned}
s &= \overline{\text{ss}} + \frac{\overline{\text{Ss}}}{2} \\
&= (\overline{\text{M.ss}} + \overline{\text{N.ss}} + \overline{\text{MN.ss}}) + \frac{\overline{\text{M.}\qquad} + \overline{\text{N.Ss}} + \overline{\text{MN.Ss}}}{2}
\end{aligned}
\tag{41}
$$

After the gene frequencies have been estimated from the distribution of the phenotypes in the population, the expected frequencies of the phenotypes can be recalculated with the aid of table 15 or table 16. Then, the chi square test can be applied as a test of the genetic theory. This has been done by Fisher,[93] using maximum likelihood estimates of the gene frequencies, which differ only slightly from the estimates obtained with the simpler method (cf. Wiener[92]) described here. For example, in one such test, a chi square value of 1.74 was obtained with 2 degrees of freedom, indicating a satisfactory agreement with the genetic theory.

Family Investigations. When the factor **S** is considered by itself, family investigations support the hypothesis of heredity as a simple Mendelian dominant. For example, Race and Sanger[94] have compiled 412 families with 941 children, and in 97 families where both parents were type s, all 221 children were likewise type s.

In family studies, when the bloods of all the individuals are tested for factors **M**, **N** and **S** simultaneously, the relationship between them becomes apparent immediately. In family 1 of table 17, the parents are type MN and type M, respectively, and, as expected, about half of the children are type MN and half are type M. Also, the parents are type S and type s, respectively, and the former is evidently heterozygous, since three of the children are type S and four are type s. However, while the type M parent lacked the factor **S** the situation is reversed in the children; that is, the type M children have the factor **S**. The reason for this becomes apparent when the phenotype of each individual is converted into the appropriate genotype, in conformity with the genetic theory. The same method of analysis is readily applicable to other families in the table, and in published studies on the subject.

Significantly when in a family factor **S** is associated with blood factor **M** in a parent, it is similarly associated in his children and in other members of his family. The same holds for factor **S** when it is

TABLE 17

ILLUSTRATIVE FAMILIES DEMONSTRATING THE RELATIONSHIP BETWEEN FACTOR **S** AND THE THREE M–N TYPES

(After Race and Sanger: *Blood Groups in Man*, 2nd ed., p. 68)

Family Number	Mating	Children						
		1	2	3	4	5	6	7
1.	MN.S × M.s	M.S	MN.s	MN.s	MN.s	M.S	MN.s	M.S
	$L^{S}l \times LL$	$L^{S}L$	Ll	Ll	Ll	$L^{S}L$	Ll	$L^{S}L$
2.	N.s × MN.S	MN.s	N.S	MN.s				
	$ll \times Ll^{S}$	Ll	$l^{S}l$	Ll				
3.	MN.S × N.s	MN.s	MN.s	N.S				
	$Ll^{S} \times ll$	Ll	Ll	$L^{S}l$				
4.	MN.S × MN.s	MN.S	N.s	MN.s				
	$L^{S}l \times Ll$	$L^{S}l$	ll	Ll				

associated with the factor **N**. To date, no evidence of crossing over between M/N and S/s has been found in family studies. This, in our opinion, indicates that factors **S** and **M** are not properties of different substances, but of one and the same agglutinogen, and thus supports the theory of multiple alleles in preference to that of linked genes.

Factor U. Wiener, Unger and Gordon,[95, 96] in 1954, encountered a new antibody in the serum of a woman who had a fatal hemolytic transfusion reaction. The factor it recognized was designated as **U** because of its almost universal distribution, as follows:

	Positive	Negative
Negroids	977	12
Caucasoids	1,100	0

The investigators also studied a family in which the father and two children lacked the factor while the mother and one child possessed the factor. This and other results supported the hypothesis that the blood factor is transmitted as a simple Mendelian dominant. It was noticed that the twelve individuals who lacked the factor **U** were all of type N or type MN, suggesting some relationship to the M-N types. Subsequently, in studies carried out with a second anti-**U** serum, Greenwalt et al.[97] confirmed these observations, and noted that all individuals lacking factor **U** also lacked both factors **S** and **s**. In this

TABLE 18
THE M–N–S–U SERIES OF ALLELIC GENES

Gene	Approximate Frequency (Per Cent) Among		Corresponding Agglutinogen	Blood Factors Present*
	Caucasoids	Negroids†		
L^S	24.7	8.7	M.s	**M**, **S**, and **U** (and 𝔑)‡
L	28.3	40.2	M.s	**M**, **s**, and **U** (and 𝔑)‡
L^u	Extremely rare		M.u	**M** (and 𝔑)‡
l^S	8.1	7.4	N.S	**N**, **S**, and **U**
l	38.9	43.7	N.s	**N**, **s**, and **U**
l^u	Extremely rare		N.u	**N**

* For simplicity in this chart the factors $\mathbf{M}_i$, $\mathbf{M}_{ii}$, $\mathbf{M}_{iii}$,. . . .$\mathbf{N}_i$, $\mathbf{N}_{ii}$, etc. are not listed (cf. page 46).

† Based on the data of Miller, Rosenfield, and Vogel, (*Amer. J. Phys. Anthrop.*, **9:** 115, 1951).

‡ Agglutinogen M reacts weakly with anti-**N** serum. This, by definition, means that the agglutinogen M has an N-like factor, which may be designated 𝔑.

way, the series of alleles in the M-N system was increased to six as shown in table 18.

Other Complexities of the M-N-S System

In 1934, Landsteiner, Strutton and Chase[98] immunized rabbits with blood of a Negroid (Mr. Hunter) and obtained an antiserum which agglutinated the blood of 7.3 per cent of Negroids and of 0.5 per cent of Whites. Wiener[1b] also tested this serum and noticed that all blood giving distinct positive reactions were type N or type MN, suggesting a relationship to the M-N system.

Chalmers, Ikin and Mourant,[99] in an attempt to repeat these studies, obtained an antiserum, prepared by injecting the blood of a Nigerian (Mr. Henshaw) into a rabbit. In tests with this antiserum (anti-**He**), 38 among 1,390 West Africans, or 2.7 per cent, gave positive reactions, while no positive reactions were obtained with the blood cells of 1,500 Europeans. Comparative tests with the original anti-**Hu** serum, some of which was still available, showed anti-**Hu** and anti-**He** to be different. With fresh blood from the original donor (Mr. Hunter) rabbits were immunized, and with the newly prepared antiserum Chalmers et al., and Shapiro[100] carried these studies further. Family studies showed that factors **He** and **Hu** are related to the properties

M and N in a manner similar to that of factors **S**, **s**, and **U**. For example, when **Hu** or **He** were associated with **M** in the parents there was a similar association in the children. This again indicates that **Hu** and **He** do not represent attributes of separate substances, but together with **M**, **N**, **S**, **s**, and **U**, as the case may be, are factors of one and the same agglutinogen. Because of these observations, the series of allelic genes for this system was extended to 24 by Shapiro, in order to account for the results of his studies in which the simultaneous heredity of **M**, **N**, **S**, **s**, **U**, **Hu**, and **He** was examined.

Additional complexities were added by the reports of two closely related blood factors designated **Gr** and **Mi**^{a}, which have been shown to be similarly associated with the M-N-S system.[101, 102, 103] The most interesting of the recent developments is Allen's finding[104] of a potent serum for an extremely rare blood factor which he showed to be related to the M-N system. He reported a pedigree in which this blood factor occurred in 7 members of the family. In this family he encountered two apparent contradictions to the heredity of the three M-N types, namely, two brothers, both type N and each with a type M child. The paradox could be explained by postulating a new extremely rare allele which produced an agglutinogen having the newly discovered blood factor but lacking both the blood factors **M** and **N**. If the allele is designated l^g, then the type N father's genotype could be ll^g while the type M child's genotype would be Ll^g.

CHAPTER IV

The P System

In 1927, when Landsteiner and Levine prepared the first anti-**M** and anti-**N** rabbit immune serums, they also obtained an antiserum which detected still another property of human blood, that was designated by them as **P**. Subsequently they found the same antibody, anti-**P**, in the serums of occasional normal human beings, and in animals (horse, pig, rabbits),[105] and it has since been found that the same antibody can occasionally result from isoimmunization in response to blood tranfusion.[106] In 1936, Furuhata and Imamura[107] described a blood factor which they designated as **Q**, detected with the aid of absorbed pig serum. Japanese investigators have carried out many studies on factor **Q**, but since comparative tests have shown **P** and **Q** to be the same, the original symbol **P** will be used exclusively in this discussion. It is of interest that among Japanese, the frequency of the **P** (or **Q**) factor is only approximately 30 per cent, in contrast to the frequency of 79 per cent among Caucasoids, and 98 per cent in Negroids.[108]

Family studies have shown the factor **P** to be inherited as a simple Mendelian dominant (cf. table 19).

Almost all anti-**P** reagents available up to now have had the properties of cold agglutinins. Moreover, such serums, whether from human beings or animals, generally contain antibodies reactive for all human red cells. Attempts to remove the non-specific antibodies by absorption tend to weaken the anti-**P** agglutinins as well. In addition, positively reacting cells from different individuals do not react with equal intensity, and all gradations from strong agglutination to weak clumping may be observed. Thus, testing for the **P** factor presents difficulties similar to those of subgrouping blood of groups A and AB.

In a review on the origin of naturally occurring hemagglutinins, Wiener[108a] wrote: "Since anti-**P** cold agglutinins occur not infrequently in normal human serum, . . . the present author hazards the prediction that future studies on bacteria may reveal them to contain antigens related to agglutinogen P in specificity." In fact, among 132 cases of hydatid disease of whom about 25 might be expected to **P** negative, Cameron and Staveley[108b] found 2 with strong anti-**P** agglutinins. Of 25 cyst fluids, 19 inhibited anti-**P** serum. In only 6 of the 25, where the scolices were absent or degenerated, did inhibition fail to occur. This confirms Wiener's theory of the heterogenetic origin of cold anti-**P** agglutinins, the antigenic stimulus apparently being provided by the P-like substance present in echinococcus cyst fluids containing scolices.

From the incidence of the **P** factor in the population, the gene frequencies are readily calculated. Thus, among Caucasoids, $p = \sqrt{(P-)} = \sqrt{0.21} = 0.46$ or 46 per cent, and $P = 1 - p = 0.54$ or 54 per cent. Similarly, among Negroids, $p = \sqrt{0.02} = 0.14$ or 14 per cent, and $P = 0.86$ or 86 per cent; while, among Chinese, $p = \sqrt{(0.70)} = 0.84$ or 84 per cent, and $P = 0.16$ or 16 per cent.

Factor Tj^a. In 1951, Levine et al.[109] encountered in the serum of a woman an antibody which agglutinated the red cells of 3,000 consecutive random group O individuals, but did not react with the red blood cells of the donor herself or those of one of her three siblings. The factor detected by this serum was named the **Jay** or **Tj**^{a} factor after the donor, Mrs. Jay. Since 1951, in eight additional investigations carried out in South Africa, Australia, Japan, the United States, and other parts of the world, antiserums of identical specificity

TABLE 19
HEREDITY OF THE P AGGLUTINOGEN
(After Dahr and Zehner)

Mating	Number of Families	Number of Children		
		P+	P−	Totals
P+ × P+	249	677	79	756
P+ × P−	134	286	179	465
P− × P−	34	(4)	94	98
Totals................	417	967	352	1319

TABLE 20
THE P SERIES OF ALLELIC GENES

Genes	Corresponding Agglutinogens	Blood Factors
P^1	P_1	**P, T$_j$^{a}**
P^2	P_2	**T$_j$^{a}**
p	p	None demonstrated to date

were encountered. In a number of these studies the **Tj**^{a}-negative type was found in more than one member of the family. It is of interest that many of the **Tj**^{a}-negative individuals had the **Tj**^{a} antibody in their serums.

If one assumes that the Tj type is determined by a pair of allelic genes Tj^a and Tj^b, then Tj(a−) individuals would all be of genotype Tj^bTj^b. The frequency of gene $Tj^b = \sqrt{\text{Tj(a—)}}$, or less than $\sqrt{0.00033}$ $= 0.018$ or 1.8 per cent, while the frequency of gene Tj^a is greater than 0.982 or 98.2 per cent.

Sanger[110] tested 15 Tj(a−) individuals from 9 families and found that they all were also **P** negative, which suggested a relationship between the **Tj**^{a} factor and the P system. In fact she found that when anti-**Tj**^{a} serum was absorbed with **P**-negative blood until it no longer reacted with the absorbing blood, an antibody fraction remained in the serum that gave reactions corresponding to anti-**P**. To account for these findings, Sanger has postulated the existence of three allelic genes P^1, P^2 and p at the P locus. The P^1 gene is postulated to determine the agglutinogen found in the type previously classified as **P**-positive; P^2 determines the agglutinogen of **P**-negative blood, while p determines the agglutinogen of Tj(a−) blood. These concepts are summarized in table 20.

CHAPTER V

The Rh-Hr System

The Rh-Hr system is the most complicated, and, second to the A-B-O system, is the most important of the human blood systems, by virtue of its clinical implications. The past fifteen years have witnessed a phenomenal advance in the knowledge of blood group serology and genetics, which was initiated by the discovery of the Rh-Hr types.

The Rh_0 Factor

Of the Rh-Hr factors the factor $\mathbf{Rh_0}$ holds a position of special prominence. It is of both theoretical and historical interest that its discovery came about in an indirect manner, with the use of an antiserum prepared in rabbits against rhesus monkey blood.[111] The $\mathbf{Rh_0}$ factor is the most common source of clinical symptoms because, as demonstrated by experimental immunization studies carried out in human volunteers, it is the most antigenic of the Rh-Hr factors.[112] It appears to represent a special structure within the Rh-Hr agglutinogen, since red cells can be coated with $\mathbf{Rh_0}$ blocking antibody without interfering with the reactions of the red cells with other antibodies such as anti-**rh′**, anti-**rh″**, and anti-**hr′**. To indicate the special position of the $\mathbf{Rh_0}$ factor, therefore, the symbol of this factor has been assigned a capital "R", while all of the other Rh factors have a small "r". The subscript of the symbol "$\mathbf{Rh_0}$" also sets it apart from the other Rh-Hr factors.

In the first study on the heredity of the Rh-Hr types, tests could be made only for the factor $\mathbf{Rh_0}$. In this investigation, carried out by Landsteiner and Wiener[113] in 1941, the antiserums prepared in guinea-pigs against rhesus monkey blood as well as human anti-$\mathbf{Rh_0}$ serum were used.

Among 448 persons tested with these serums, the following distribution was found.

	Number	Per Cent
Rh positive	379	84.6
Rh negative	69	15.4

Under the hypothesis that the $\mathbf{Rh_0}$ factor is inherited as a simple Mendelian dominant, the gene frequencies are readily estimated as follows:

$$rh = \sqrt{\text{Rh negative}} = \sqrt{0.154} = 39.2 \text{ per cent}$$

$$Rh = 1 - rh = 60.8 \text{ per cent}$$

The expected frequencies of the three theoretically possible genotypes can then be readily calculated.

$$\text{Rh positive} \begin{cases} \text{genotype } RhRh = (0.608)^2 = 37.0 \text{ per cent} \\ \text{genotype } Rhrh = 2(0.608 \times 0.392) = 47.6 \text{ per cent} \end{cases}$$

$$\text{Rh negative; genotype } rhrh = (0.392)^2 = 15.4 \text{ per cent}$$

Family Studies. Landsteiner and Wiener tested 60 families with 237 children for the $\mathbf{Rh_0}$ factor, and in table 21 their findings are summarized. When the distribution the $\mathbf{Rh_0}$ factor is compared with that expected under the genetic theory the findings prove to be in satisfactory agreement.

Landsteiner and Wiener also showed that the $\mathbf{Rh_0}$ factor is not sex linked and excluded any relation with the A-B-O and M-N systems.

Serology. While in the original investigations, anti-rhesus immune serum was used, it was subsequently found that antiserums of greater avidity and higher titer are more readily obtained and in larger amounts from mothers of erythroblastotic babies as well as from deliberately immunized donors. Therefore, anti-rhesus serums are no longer used, and remain only of theoretical and historical im-

TABLE 21

HEREDITARY TRANSMISSION OF THE $\mathbf{Rh_0}$ FACTOR

(From Landsteiner and Wiener, 1941)

Mating	Number of Families	Number of Children		
		Rh+	Rh−	Totals
Rh+ × Rh+	42	151	7	158
Rh+ × Rh−	12	37	11	48
Rh− × Rh−	6	0	31	31
Totals	60	188	49	237

portance. It is interesting that the human anti-$\mathbf{Rh}_0$ serum, even those of maximum titer and avidity, give little or no reaction with red cells of rhesus monkeys. This seemingly paradoxical observation is another example of so-called non-reciprocal reactions, and is inexplicable under the hypothesis of a strict one-to-one correspondence between antigen and antibody.

It was early observed,[114] in many Rh-negative patients with clinical manifestations of isosensitization, that Rh antibodies could not be demonstrated by the classic agglutination method in saline media. This led Wiener in 1941 to postulate that Rh antibodies in particular (and all antibodies in general), exist in two molecular forms, one of which was capable of combining with Rh-positive cells in saline media but without clumping the cells. In 1944, when suitable human anti-Rh agglutinating serum was more easily available, Wiener[115] devised the blocking test to demonstrate the second form of antibody. Later on, Race[116] and Diamond[117] independently encountered the $\mathbf{Rh}_0$ blocking antibody when they tried unsuccessfully to produce a polyvalent anti-$\mathbf{Rh}_0{}'$ serum by pooling anti-$\mathbf{Rh}_0$ and anti-$\mathbf{rh}'$ serums. The antibody which clumps cells in saline media is variously known as "bivalent", agglutinating, or "complete" antibody, while the antibody detected by the blocking test has been designated "univalent", blocking, conglutinating, or "incomplete" antibody.

The original blocking test detected univalent Rh antibodies only when present in high titer, and for this test a satisfactory agglutinating anti-Rh serum is also needed. Fortunately, other more sensitive and more practicable tests for univalent antibodies have been devised, notably, the conglutination[118] and albumin tests,[119] the anti-globulin test,[120] and the use of test cells treated with proteolytic enzymes.[121] Since univalent antibodies are encountered far more frequently than bivalent antibodies, these new tests resulting from the discovery of the Rh factor, opened up the entire field of blood grouping, and may be credited with the phenomenal advances of the past 15 years.

Factor *rh′*

In 1941, Wiener[114] encountered a patient who had a hemolytic transfusion reaction and whose serum contained an antibody which reacted with 70 per cent of Caucasoid individuals, in contrast with the original anti-Rh serum which agglutinated the red cells of 85 per cent of such persons (cf. table 22). By the use of a 2 × 2 contingency table, it was possible to show that the factor demonstrated by this serum was independent of the A-B-O, M-N and P systems, but related to the **Rh**$_0$ factor. The problem then arose as to the nature of the relationship between the new blood factor, which is to-day known as **rh′**, and the original **Rh**$_0$ factor.

It is instructive as well as of historical interest to give the reasoning as originally presented in the 1943 edition of Wiener's book, Blood Groups and Transfusion, making necessary changes to correspond to modern terminology, in order to avoid misunderstanding. The quotation modified in this way reads as follows.*

"The reactions could be explained most simply by postulating the existence of two agglutinogens Rh$_0$ and rh′ corresponding to the agglutinins anti-**Rh**$_0$ and anti-**rh′**. The four sorts of blood would then have the compositions Rh$_0$rh′, Rh$_0$, rh′, and Rh negative, respectively. This assumption would imply the existence, however, of two corresponding genes R^0 and r', which would have to be either independent, linked or allelic. The first two possibilities are excluded since the product of the frequencies Rh$_0$rh′ × Rh negative is much greater than Rh$_0$ × rh′. Moreover the existence of allelic genes R^0 and r' in individuals whose blood reacts with both anti-**Rh**$_0$ and anti-**rh′** would necessitate that this class not exceed 50 per cent,† while the actual frequency is 70 per cent. Accordingly, the observations are best explained by assuming the existence of 3 qualitatively different Rh agglutinogens instead of only 2, one type reacting with anti-**Rh**$_0$ serum but not anti-**rh′**, a second reacting with anti-**rh′** but not with anti-**Rh**$_0$, and a third reacting with both sorts of anti-Rh serums."

At this stage, the genetics of the Rh-Hr system could be summarized as follows. Four allelic genes are postulated. Gene r gives rise to an agglutinogen lacking both factors **Rh**$_0$ and **rh′**; gene r' corresponds to an agglutinogen having only factor **rh′**; similarly gene R^0 controls an agglutinogen having only factor **Rh**$_0$, and finally gene R^1 deter-

* In the original quotation the anti-**Rh**$_0$ serum was designated anti-**Rh**$_1$, while anti-**rh′** was designated as anti-**Rh**$_2$.

† On page 45, when discussing this question, it was also pointed out that the frequencies of type MN, and of group AB similarly could not exceed 50 per cent in a population at equilibrium.

TABLE 22

RELATIONSHIP AMONG THE REACTIONS OF DIFFERENT ANTI-Rh SERUMS

Antiserums	Originally Described by	Present Designation	Approximate Per Cent Positive	Reactions of Various Bloods among White Individuals			
				About 70 %	About 15%	About 2%	About 13%
Type 1 (standard anti-Rh)	Landsteiner and Wiener	Anti-$\mathbf{Rh}_0$	85	Pos.	Pos.	Neg.	Neg.
Type 2	Wiener and Landsteiner	Anti-**rh′**	72	Pos.	Neg.	Pos.	Neg.
Type 3	Levine, Burnham, Katzin and Vogel	Anti-$\mathbf{Rh}_0'$	87	Pos.	Pos.	Pos.	Neg.

TABLE 23

THE FOUR PHENOTYPES DETERMINED BY FACTORS $\mathbf{Rh}_0$ AND **rh′** AND THEIR TEN THEORETICALLY POSSIBLE GENOTYPES

Phenotype	Reaction with Serum		Possible Genotypes
	Anti-$\mathbf{Rh}_0$	Anti-**rh′**	
rh	−	−	*rr*
rh′	−	+	$r'r$ and $r'r'$
Rh_0	+	−	R^0r and R^0R^0
Rh_1	+	+	R^1r, R^1R^0, R^0r', R^1R^1, and R^1r'

mines an agglutinogen characterized by both factors $\mathbf{Rh}_0$ and **rh′**. Then, corresponding to the four phenotypes there could theoretically be 10 possible genotypes as shown in table 23.

Wiener and Landsteiner[122] studied the simultaneous heredity of factors $\mathbf{Rh}_0$ and **rh′** in 47 families with 133 children. In this small series, unfortunately, there appeared no instance of the rarest of the four blood types, namely, type rh′. However, the results, as far as they went, were in full accord with the genetic theory.

Eight Rh Types

Factor rh″. Early in 1943, Wiener[123] described still another antiserum which agglutinated the red cells of only 30 per cent of the white population. Statistical analysis proved that this factor was related to the Rh-Hr system, and to indicate that it is now known as **rh″**. Considering only the reactions of anti-**rh′** and anti-**rh″** serums, four types of blood are possible which could be designated as rh, rh′, rh″, and rh′rh″, respectively.* The distribution of these four types among 280 individuals proved to be as follows: type rh, 19.4 per cent; type rh′, 50.3 per cent; type rh″, 12.8 per cent; and type rh′rh″ 17.5 per cent. Since the product rh′ × rh″ is considerably greater than rh × rh′rh″ it is clear that rh′ and rh″ are not independent. A tentative genetic theory was proposed by Wiener,[124] that the factors **rh′** and **rh″** are transmitted by triple allelic genes in a manner comparable to the heredity of the four A-B-O blood groups. Based on this hypothesis, the gene frequencies were readily estimated as follows.†

$$r = \sqrt{\text{rh}} = \sqrt{0.194} = 44.5 \text{ per cent}$$

$$r' = \sqrt{\text{rh} + \text{rh}'} - \sqrt{\text{rh}} = \sqrt{0.697} - \sqrt{0.194} = 39 \text{ per cent}$$

$$r'' = \sqrt{\text{rh} + \text{rh}''} - \sqrt{\text{rh}} = \sqrt{0.322} - \sqrt{0.194} = 12 \text{ per cent}$$

So that, $r + r' + r'' = 95.5$ per cent.

Thus, the agreement with the genetic theory was not entirely satisfactory, since the sum of the estimated gene frequencies fell short of 100 per cent, and the theoretical frequency of type rh′rh″, which should equal $2 \times r' \times r''$ or about 9.5 per cent, was considerably less that the observed frequency. These discrepancies were correctly attributed to difficulties in technique, due to the poor quality of the antiserums available at that time. In addition, no allowance was made for additional genes giving rise to agglutinogens with both factors **rh′** and **rh″**. Such genes are now known to exist, but their incidence in the population is so low that failure to take them into account must have been only a minor contributing factor to the deviation found in this early study.

The Eight Rh Phenotypes. When tests are made with all three antiserums simultaneously, eight types of blood can be distinguished

* In the original paper, the four types were designated as Rh negative, Rh_1 Rh_2, and Rh_1Rh_2, respectively.

† The symbols for the genes as used here of necessity have a different meaning from the same symbols when used in connection with the calculations that include factor $\mathbf{Rh_0}$, as will be seen shortly.

TABLE 24

SCHEME OF THE EIGHT Rh TYPES

(Wiener, A. S., *Dade County M. A. Bull.*, 1949)

Rh-Negative Types				Rh-Positive Types			
Destination of types	Reaction with serum			Designation of types	Reaction with serum		
	Anti-**rh'**	Anti-**rh''**	Anti-$\mathbf{Rh_0}$		Anti-**rh'**	Anti-**rh''**	Anti-$\mathbf{Rh_0}$
rh	−	−	−	Rh_0	−	−	+
rh'	+	−	−	Rh_1	+	−	+
rh''	−	+	−	Rh_2	−	+	+
rh'rh''	+	+	−	Rh_1Rh_2	+	+	+

The types are named after the Rh blood factors (and agglutinogens) present in the red cells, Rh_1 being short for Rh_0rh' (or Rh_0') and Rh_2 short for Rh_0rh'' (or Rh_0''). Similarly, it is often convenient to use rh_y in place of rh'rh'' and Rh_z in place of Rh_1Rh_2 (or $Rh_0'Rh_0''$).

as shown in table 24. These eight phenotypes are easy to remember if they are arranged in the form of a double scheme of four types each, analogous to the four A-B-O blood groups.

The distribution of the types is different in diverse ethnic groups, and the results of few representative studies are shown in table 25. As can be seen from the table, Caucasoids have the highest frequency of type rh, while this type is virtually absent in Mongoloids. The most striking feature of negroid populations is the very high frequency of type Rh_0.

Genetic Theory. To account for the existence of the eight phenotypes, Wiener[124] found it necessary to postulate the existence of at least six allelic genes as shown in table 26. Under this concept there are 21 theoretically possible genotypes corresponding to the eight possible phenotypes as shown in table 27.

Gene Frequency Analysis. The theory of multiple alleles to account for the eight Rh types has been tested by gene frequency analysis and by family studies. With the aid of table 27, the following formulae for the frequencies of the genes in terms of the frequencies of Rh phenotypes can readily be derived.

$$r = \sqrt{\text{rh}} \qquad (31)$$

$$r' = \sqrt{\text{rh}' + \text{rh}} - \sqrt{\text{rh}} \qquad (32)$$

$$r'' = \sqrt{\text{rh}'' + \text{rh}} - \sqrt{\text{rh}} \qquad (33)$$

TABLE 25
REPRESENTATIVE DATA ON THE RACIAL DISTRIBUTION OF THE EIGHT Rh TYPES

Ethnic Group	Approximate Frequency (Per Cent) of the Blood Types							
	rh	rh′	rh″	rh′rh″	Rh_0	Rh_1	Rh_2	Rh_1Rh_2
Caucasoids (N.Y.C.)[1]	13.5	1.0	0.5	.02	2.5	53.0	15.0	14.5
Negroids								
N.Y.C.[2]	7.5	1.5	0	0	45.0	25.0	15.5	5.5
Africa[3]	3.75	0.75	0	0	70.0	15.0	9.0	1.5
Puerto Ricans[4]	10.1	1.7	0.5	0	15.1	39.1	19.6	14.0
Chinese[5]	1.5	0	0	0	0.9	60.6	3.0	34.4
Japanese[6]	0.6	0	0	0	0	51.7	8.3	39.4
Filipinos[7]	0	0	0	0	0	87.0	2.0	11.0
Mexican Indians[8]	0	0	0	0	1.1	48.1	9.5	41.2

(Modified after Wiener A. S., *Am. J. Clin. Path.*, **16:** 477, 1946)

[1] Wiener et al.; Unger et al.; Levine
[2] Wiener et al.; Levine
[3] Hubinont et al.
[4] Torregrosa
[5] Wiener et al.
[6] Waller and Levine; Miller and Taguchi
[7] Simmons and Graydon
[8] Wiener et al.

$$R^0 = \sqrt{\mathrm{Rh_0 + rh}} - \sqrt{\mathrm{rh}} \tag{34}$$

$$R^1 = \sqrt{\mathrm{Rh_1 + rh' + Rh_0 + rh}} - \sqrt{\mathrm{rh' + rh}} - \sqrt{\mathrm{Rh_0 + rh}} + \sqrt{\mathrm{rh}} \tag{35}$$

$$R^2 = \sqrt{\mathrm{Rh_2 + rh'' + Rh_0 + rh}} - \sqrt{\mathrm{rh'' + rh}} - \sqrt{\mathrm{Rh_0 + rh}} + \sqrt{\mathrm{rh}} \tag{36}$$

As an example the formulae will be applied to the figures given for Caucasoids in table 25.

$$r = \sqrt{0.135} = 36.7 \text{ per cent}$$

$$r' = \sqrt{0.010 + 0.135} - \sqrt{0.135} = 1.4 \text{ per cent}$$

$$r'' = \sqrt{0.005 + 0.135} - \sqrt{0.135} = 0.7 \text{ per cent}$$

$$R^0 = \sqrt{0.025 + 0.135} - \sqrt{0.135} = 3.3 \text{ per cent}$$

$$R^1 = \sqrt{0.530 + 0.010 + 0.025 + 0.135} - \sqrt{0.010 + 0.135} - \sqrt{0.025 + 0.135} + \sqrt{0.135} = 42.3 \text{ per cent}$$

TABLE 26

SIX ALLELIC GENES AND THEIR CORRESPONDING AGGLUTINOGENS

Gene	Agglutinogen	Blood Factors Present
r	rh	None
r'	rh′	**rh′**
r''	rh″	**rh″**
R^0	Rh_0	$\mathbf{Rh}_0$
R^1	Rh_1	$\mathbf{Rh}_0$ and **rh′**
R^2	Rh_2	$\mathbf{Rh}_0$ and **rh″**

TABLE 27

THE SIX ALLELE THEORY OF INHERITANCE OF THE Rh TYPES

Phenotypes	Corresponding Genotypes
rh	rr
rh′	$r'r'$ and $r'r$
rh″	$r''r''$ and $r''r$
rh′rh″	$r'r''$
Rh_0	R^0R^0 and R^0r
Rh_1	R^1R^1, R^1r', R^1r, R^1R^0, and R^0r'
Rh_2	R^2R^2, R^2r'', R^2r, R^2R^0, and R^0r''
$\mathrm{Rh}_1\mathrm{Rh}_2$	R^1R^2, R^1r'', and R^2r'

$$R^2 = \sqrt{0.150 + 0.005 + 0.025 + 0.135} - \sqrt{0.005 + 0.135} - \sqrt{0.025 + 0.135} + \sqrt{0.135} = 15.4 \text{ per cent}$$

The sum of these estimated gene frequencies comes to 99.8 per cent, which is a very satisfactory agreement with the genetic theory. Actually, the sum of the gene frequencies should fall short of 100 per cent, because of the existence of additional alleles as will be explained later on. However, these additional genes are so rare that they would have no noticeable effect on the calculations. In the early studies, seven of the eight theoretically possible phenotypes were encountered, that is, all but the rare type rh′rh″. Why this type was not found at first was easy to understand from the genetic theory, since its expected frequency was only $2 \times r' \times r''$, or $2 \times 0.014 \times 0.007 = 0.000196$ or about 1 in 5,000. The subsequent

TABLE 28

AUTHOR'S STUDIES (1943–1953) ON THE HEREDITY OF THE EIGHT Rh TYPES

Mating	Number of Families	Number of Children of Type								Totals
		rh	rh′	rh″	rh′rh″	Rh_0	Rh_1	Rh_2	Rh_1Rh_2	
rh × rh	32	53	0	0	0	0	0	0	0	53
rh × rh′	3	1	2	0	0	0	0	(1)	0	4
rh × rh″	2	2	0	5	0	0	0	0	0	7
rh × rh′rh″	1	1	0	0	1	0	0	0	0	2
rh × Rh_0	31	15	0	0	0	33	0	0	0	48
rh × Rh_1	599	182	5	0	0	40	731	(1)	0	959
rh × Rh_2	126	72	0	0	0	4	0	144	0	220
rh × Rh_1Rh_2	152	3	1	0	0	0	112	120	7	243
rh′ × Rh_0	3	0	2	0	0	1	2	0	0	5
rh′ × Rh_1	31	7	7	0	0	5	30	0	0	49
rh′ × Rh_2	8	2	5	0	0	0	0	3	3	13
rh′ × Rh_1Rh_2	10	0	0	0	0	0	6	4	7	17
rh″ × Rh_1	14	3	0	1	0	0	8	0	11	23
rh″ × Rh_2	3	0	0	2	0	0	0	3	0	5
rh″ × Rh_1Rh_2	4	0	0	0	0	0	2	0	2	4
rh′rh″ × Rh_0	1	0	0	3	0	0	0	0	0	3
rh′rh″ × Rh_1	2	0	0	0	0	0	1	0	2	3
rh′rh″ × Rh_2	1	0	0	0	0	0	0	1	1	2
Rh_0 × Rh_0	1	0	0	0	0	1	0	0	0	1
Rh_0 × Rh_1	19	6	1	0	0	4	16	0	0	27
Rh_0 × Rh_2	4	1	0	0	0	4	0	2	0	7
Rh_0 × Rh_1Rh_2	3	0	0	0	0	0	2	1	0	3
Rh_1 × Rh_1	125	12	1	0	0	6	188	0	0	207
Rh_1 × Rh_2	57	12	4	0	0	5	35	13	29	98
Rh_1 × Rh_1Rh_2	71	0	1	0	0	0	74	17	53	145
Rh_2 × Rh_2	10	3	0	0	0	0	0	13	0	16
Rh_2 × Rh_1Rh_2	18	0	0	0	0	0	8	17	12	37
Rh_1Rh_2 × Rh_1Rh_2	14	0	0	0	0	0	6	4	8	18
Totals............	1345	375	29	11	1	103	1221	344	135	2219

The numbers in parenthesis represent contradictions to the genetic theory, apparently due to illegitimacy.

The excess of families with Rh-negative parents is explained by the fact that this material was largely compiled from families confronted with the problem of Rh sensitization.

discovery of type rh′rh″ is one of the numerous facts which have been unearthed, supporting the theory of multiple alleles.

Family Studies. Further evidence of the validity of the theory of multiple alleles was obtained by family investigations.[125, 126] In table 28, for example, are summarized the studies of Wiener and his associates during the period 1943 to 1953 on the heredity of the eight Rh types. Similar studies have been done by Race et al., and additional data have been accumulated more recently in investigations not yet published. With only two exceptions, which may be ascribed to illegitimacy, the data obtained on the 2,119 children shown in table 28 support the theory of multiple alleles.

Actually, a few other exceptions were encountered in certain families, where one of the parents belonged to type rh′rh″ or to type Rh_1Rh_2. These apparent exceptions to the genetic theory could be explained by postulating the existence of two additional alleles in the series of allelic genes.[127, 128] Thus, under the theory of six allelic genes matings with parents rh′rh″ × rh would have to be of genotypes $r'r'' \times rr$, so that half of the children would be expected to be type rh′ and half to be type rh″. Actually, in some of these families half of the children were type rh′rh″ while half were type rh. By postulating the existence of an additional allelic gene r^y, determining a corresponding agglutinogen rh_y having the blood factors **rh′** and **rh″** but lacking factor $\mathbf{Rh}_0$, it was possible to account for these exceptional families. Similarly, it was necessary to postulate the existence of a gene R^z determining a corresponding agglutinogen Rh_Z having all three factors $\mathbf{Rh}_0$, **rh′**, and **rh″**, in order to account for anomalies in certain families with one or both parents of type Rh_1Rh_2. In table 29, are shown three families illustrating the transmission of the rare genes r^y and R^z.

Serology. For identifying the 8 Rh types, it is necessary to have available antiserums of specificities anti-$\mathbf{Rh}_0$, anti-**rh′**, and anti-**rh″**, respectively. Pure anti-**rh′** serum can be obtained from type Rh_2 individuals sensitized against type Rh_1 blood, and, similarly, pure anti-**rh″** serum is obtained from type Rh_1 individuals sensitized against type Rh_2 blood. Unfortunately, such monovalent antiserums are rare and the reagents available commercially are mostly derived from type rh individuals sensitized to type Rh_1 or type Rh_2 blood, respectively. Thus, the anti-**rh′** reagents most commonly used are really anti-$\mathbf{Rh}_0'$ serums to which anti-$\mathbf{Rh}_0$ blocker has been added to

TABLE 29

FAMILIES ILLUSTRATING THE TRANSMISSION OF THE RARE GENES R^z AND r^y

Mating	Children				
	1	2	3	4	5
1. Rh_1Rh_2 × rh	rh	Rh_1Rh_2			
$R^zr \times rr$	rr	R^zr			
2. rh′rh″ × rh	rh	rh′rh″			
$r^yr \times rr$	rr	r^yr			
3. Rh_1Rh_2 × rh	rh	rh	Rh_1Rh_2	rh	Rh_1Rh_2
$R^zr \times rr$	rr	rr	R^zr	rr	R^zr

prevent the anti-$\mathbf{Rh_0}$ agglutinin from giving rise to agglutination under the conditions of the test. Similarly, most anti-**rh″** reagents are anti-$\mathbf{Rh_0''}$ serums to which anti-$\mathbf{Rh_0}$ blocker has been added. If the directions for the tests are not followed scrupulously, or when dealing with highly agglutinable red cells which are difficult to block, such as cells from homozygous type Rh_2 or type Rh_0 individuals, blocking may be incomplete or may fail and the clumping resulting may be misinterpreted as indicative of the presence of the **rh′** and **rh″** factor. Such errors have caused serious misunderstandings in medicolegal cases of disputed paternity, and have also been responsible for inaccurate reports in anthropological investigations.

Hr Factors

Levine and Javert[129] encountered an antibody in the serum of an Rh-positive mother of an erythroblastotic baby which gave approximately 30 per cent positive reactions with Caucasoid bloods. Significantly, all Rh-negative bloods gave positive reactions with this serum. Because of its apparently reciprocal relation to **Rh**, they named the factor detected by this antiserum **Hr**. Independently, Race and Taylor[130] discovered an antiserum for a blood factor present in 80 per cent of Caucasoid bloods. Wiener *et al.*[131] pointed out that factor **Hr** of Levine and Javert and the factor found by Race and Taylor were in fact the same, the difference in percentage of positive reactions being due to a gene dosage effect.

More detailed study showed that the so-called **Hr** factor was related to factor **rh′** in a manner similar to the relationship between **M** and **N**, that is, individual bloods are either **rh′** positive or **Hr** positive, or both, and no blood was both **rh′** negative and **Hr** negative. To indicate this, Wiener introduced the designation **hr′** for the factor. It must be pointed out that the discovery of factor **hr′** made necessary no change in the genetic theory, since **hr′** merely defines another blood factor of agglutinogens already known and identifiable by their reactions with the three Rh antiserums alone. The table of Rh allelic genes is therefore modified only by the addition of factor **hr′** in the "blood factor" column for agglutinogens rh, rh″, Rh_0, and Rh_2 (cf. tables 26 and 31). Thus, the number of possible genotypes remains the same, but the number of possible phenotypes is increased. The additional phenotypes distinguished by the **hr′** antiserum have been designated as rh′rh for those type rh′ bloods which are positive for factor **hr′**, while those type rh′ bloods which are negative for factor **hr′** are designated as rh′rh′. Similarly, type Rh_1 bloods are subdivided by their reactions with **hr′** antiserum into type Rh_1rh for those bloods which react positively with **hr′** antiserum, and type Rh_1Rh_1 for those bloods which do not react with **hr′** antiserum.

Considering that factor **rh′** has an incidence of approximately 70 per cent in the Caucasoid population, the expected frequency of the reciprocally related factor **hr′** can be calculated with the aid of the formula:

$$\begin{aligned} \mathbf{hr'}\ \text{pos.} &= 1 - [1 - \sqrt{(\mathbf{rh'}\ \text{neg.})}]^2 \\ &= 1 - [1 - \sqrt{0.30}]^2 \\ &= 1 - (1 - 0.548)^2 \\ &= 1 - 0.204 \\ &= 0.796 \text{ or } 79.6 \text{ per cent} \end{aligned}$$

This frequency, as can be seen, is very close to the value reported by Race and Taylor. If the factors **rh′** and **hr′** are considered by themselves, three phenotypes are distinguished, the expected frequencies of which would be as follows:

$$\begin{aligned} \text{type rh}' &= \mathbf{hr'}\ \text{negative} = 21\,\% \\ \text{type hr}' &= \mathbf{rh'}\ \text{negative} = 30\,\% \\ \text{type rh}'\text{hr}' &= \text{the rest} = 49\,\% \end{aligned}$$

By using an over simplification, it is possible to account for the existence of three such types. If we postulate a pair of alleles $\overline{r'}$ and $\overline{h'}$ (the bar is placed over the "collective" gene $\overline{r'}$ to distinguish it from gene r' of the series of eight Rh alleles), there are three genotypes corresponding to the three phenotypes as follows:

Phenotype	Genotypes	Frequencies %
rh′	$\overline{r'r'}$	21
rh′hr′	$\overline{r'h'}$	49
hr′	$\overline{h'h'}$	30

It will be noted that while the incidence of the **hr′** factor is 79 per cent, among these 30 per cent are homozygous while 49 per cent are heterozygous. It is now known that blood from individuals homozygous for the **hr′** factor reacts much more avidly with anti-**hr′** serum than blood from heterozygous individuals. This is known as the gene dose effect. It is probable that the original serum of Levine and Javert was detecting only homozygous **hr′**-positive blood, which accounts for their early observations.

Fisher,[132] noticing the reciprocal relationship between factors **rh′** and **hr′**, postulated the existence of two additional Hr factors, namely, **hr″** and $\mathbf{Hr_0}$, respectively, reciprocally related to factors **rh″** and $\mathbf{Rh_0}$, respectively. As will be explained later, this prediction was associated with the genetic theory of linkage, and a different nomenclature for the Rh-Hr types. Shortly thereafter, Mourant[133] announced the discovery of an antiserum giving the reactions expected of anti-**hr″**, and additional antiserums of this specificity were found by Wiener and Peters[134] and other workers. Several announcements of the discovery of antiserums supposedly of specificity anti-$\mathbf{Hr_0}$ were also made, but these findings could not be confirmed. Even at the present time anti-$\mathbf{Hr_0}$ serum is not available, although anti-**hr′** and anti-**hr″** serums are routinely used in many laboratories specializing in Rh-Hr typing.

In table 30 are given the expected incidence of factors **hr′** and **hr″** as well as those expected of the hypothetical factor $\mathbf{Hr_0}$ among Whites, Negroes, Mongoloids and Asiatic Indians, estimated from the incidence of factors **rh′**, **rh″** and $\mathbf{Rh_0}$ in these populations, as calculated with the aid of the general formula,[135]

$$(\text{Hr}+) = 1 - [1 - \sqrt{(\text{Rh}-)}]^2 \qquad (37)$$

TABLE 30

RELATIONSHIP BETWEEN Rh AND Hr ANTISERUMS

Racial Group	Observed* Percentages of Positive Reactions with Rh Antiserums			Expected Percentages of Positive Reactions with Hr antiserums		
	rh′	**rh″**	$\mathbf{Rh_0}$	**hr′**	**hr″**	$\mathbf{Hr_0}$ (Hypopthetical)
Whites	70	30	85	80	97	63
Negroes	28	27	90	97.7	97.5	54
American Indians; Chinese and Japanese	85 to 95	40 to 60	99 to 100	43 to 63	86 to 95	0 to 20
Asiatic Indians	85	18	93	63	99.1	45

* A. S. Wiener, E. B. Sonn, and R. B. Belkin, *Proc. Soc. Exp. Biol. and Med.*, **54:** 238, 1943; A. S. Wiener, J. P. Zepeda, E. B. Sonn, and H. R. Polivka, *J. Exp. Med.*, **81:** 559, 1945.

Nomenclature. With the discovery of additional Rh-Hr factors, the relationships among the various agglutinogens become increasingly complicated, and this entails corresponding complexities in the notations for the various Rh-Hr types. When choosing the symbols for the Rh-Hr types, rational principles of notation were applied so that the resulting nomenclature is not only accurate and unambiguous, but also serves to translate the serological facts into symbols which can be easily manipulated both graphically and orally. Thus, no attempt is made to include everything that is known about a given agglutinogen in its symbol, since this would entail repeated changes in the notation with the discovery of each new fact. Rather, the symbols chosen are in the nature of easily inscribed and spoken mnemonics, which aid the memory and at the same time stress only those characteristics needed for the identification of the agglutinogen or phenotype. For example, agglutinogen rh is defined and identified by its negative reactions with antiserums $\mathbf{Rh_0}$, **rh′**, and **rh″**. With the discovery of anti-**hr′** and anti-**hr″**, no change had to be made in the symbol; instead, the definition of agglutinogen rh was merely expanded to include its reactions with these two additional antiserums. Thus, agglutinogen rh is now more fully defined as that agglutinogen which gives negative reactions with antiserums $\mathbf{Rh_0}$, **rh′**, and **rh″**,

TABLE 31

THE EIGHT "STANDARD" Rh ALLELIC GENES

Gene	Corresponding Agglutinogen	Blood Factors Present
r	rh	**hr′** and **hr″**
r'	rh′	**rh′** and **hr″**
r''	rh″	**rh″** and **hr′**
r^y	rh_y	**rh′** and **rh″**
R^0	Rh_0	**Rh_0**, **hr′**, and **hr″**
R^1	Rh_1	**Rh_0**, **rh′**, and **hr″**
R^2	Rh_2	**Rh_0**, **rh″**, and **hr′**
R^z	Rh_Z	**Rh_0**, **rh′**, and **rh″**

and positive reactions with anti-**hr′** and anti-**hr″**. As was anticipated, additional characteristics of agglutinogen rh have been discovered, notably, **hr**, and more recently, **hr^v**. It will be seen that the factor **hr** is not essential for the identification of agglutinogen rh, and therefore need not be included in its symbol. On the other hand, since factor **hr^v** identifies a property of only *certain* rh agglutinogens, it does require a distinctive symbol to indicate its presence.

In table 31 the situation resulting from the discovery of the additional two factors **hr′** and **hr″** is summarized for the 8 "standard" allelic genes. When the three Rh antiserums and the two Hr antiserums are used simultaneously, instead of only 8 Rh types a scheme of 18 Rh-Hr phenotypes results, as is shown in table 32. In order to indicate the reactions of the Hr antiserums a new convention has to be introduced, as has already been indicated. To facilitate the use of this convention the names of two of the phenotypes are changed as follows: instead of rh′rh″ the symbol rh_y is used, and instead of Rh_1Rh_2 the symbol Rh_Z is employed. In this way it is possible to indicate not only the results of the tests but also which antiserums have been used. Thus, when tests are made using only the three Rh antiserums, a single symbol is used to indicate the phenotype, while a second symbol is added when tests have been carried out for the Hr factors as well. Since all type Rh_1 bloods necessarily give positive reactions with anti-**hr″** serum,* this fact need not be indicated in the phenotype symbol. On the other hand, to indicate the reaction with anti-**hr′** the symbol Rh_1Rh_1 is used for type Rh_1 bloods which are **hr′** negative, while the symbol Rh_1rh is used for type Rh_1 bloods

* Rare exceptions to this rule will be described later.

TABLE 32

NOMENCLATURE OF THE 18 Rh-Hr BLOOD TYPES AND 36 GENOTYPES

Eight Rh Phenotypes	Reactions with Serum		Designation	
	Anti-**hr′**	Anti-**hr″**	Phenotypes	Corresponding genotypes
rh	+	+	rh	rr
rh′	+	+	rh′rh	$r'r$
	−	+	rh′rh	$r'r'$
rh″	+	+	rh″rh	$r''r$
	+	−	rh″rh″	$r''r''$
rh_y (or rh′rh″)	+	+	$rh_y rh$	$r'r''$ and $r^y r$
	−	+	$rh_y rh'$	$r^y r'$
	+	−	$rh_y rh''$	$r^y r''$
	−	−	$rh_y rh_y$	$r^y r^y$
Rh_0	+	+	Rh_0	R^0R^0 and R^0r
Rh_1	+	+	$Rh_1 rh$	R^1r, R^1R^0 and R^0r'
	−	+	Rh_1Rh_1	R^1R^1 and R^1r'
Rh_2	+	+	$Rh_2 rh$	R^2r, R^2R^0, and R^0r''
	+	−	Rh_2Rh_2	R^2R^2 and R^2r''
Rh_Z (or Rh_1Rh_2)	+	+	Rh_ZRh_0	R^1R^2, R^1r'', R^2r', R^zr, R^zR^0 and R^0r^y
	−	+	Rh_ZRh_1	R^zR^1, R^zr', and R^1r^y
	+	−	Rh_ZRh_2	R^zR^2, R^zr'', and R^2r^y
	−	−	Rh_ZRh_Z	R^zR^z and R^zr^y

which are **hr′** positive. Thus, the first half of the phenotype symbol represents the reactions obtained with the Rh antiserums, while the second half of the symbol represents the reaction with the Hr antiserums. As shown below there is generally a choice of two symbols to represent the Hr reactions:

rh or Rh_0 represents presence of both factors **hr′** and **hr″**

rh′ or Rh_1 represents absence of **hr′** and presence of **hr″**

rh″ or Rh_2 represents presence of **hr′** and absence of **hr″**

rh_y or Rh_Z represents absence of both factors **hr′** and **hr″**

As shown in table 33 the symbol actually assigned to represent the Hr reactions, in most, but not all cases makes the complete phenotype symbol correspond to the most frequent of the possible genotypes. It must be stressed that the adoption of this convention does not commit one to a specific genotype, since the symbol for phenotypes and genotypes always remain distinctive and unambiguous.

Family Studies. Studies on the heredity of the Rh-Hr blood types (136–142), support the hypothesis that **hr′** is related to **rh′**, and **hr″** is related to **rh″**, as factor **M** is related to **N**. Many studies have been carried out with anti-**hr′** serum but only a few studies have been made using anti-**hr″** serum, because the latter is less readily available. The rules of heredity can be summarized as follows:

1. Factor **hr′** or **hr″** cannot appear in the blood of a child unless it is present in one or both parents.

TABLE 33

STATISTICAL ANALYSIS OF R. R. RACE'S DATA ON THE DISTRIBUTION OF THE Rh-Hr TYPES IN LONDON

Rh-Hr Phenotypes	Number	Per Cent
rh	170	15.84
rh′rh′	0	0
rh′rh	10	0.93
rh″rh″	0	0
rh″rh	7	0.65
rh_yrh, rh_yrh', rh_yrh'', rh_yrh_y	0	0
Rh_0	19	1.77
Rh_1Rh_1	190	17.71
Rh_1rh	363	33.83
Rh_2Rh_2	29	2.70
Rh_2rh	137	12.77
Rh_ZRh_0	144	13.42
Rh_ZRh_2	4	0.37
Rh_ZRh_2, Rh_ZRh_Z	0	0

$r = 39.80$; $r' = 1.15$; $r'' = 0.81$; $R^0 = 2.19$; $R^1 = 40.57$; $R^2 = 15.28$; $R^z = 0.44$.

Thus, $\Sigma R = 100.24$.

$\sqrt{(\mathbf{rh'}-)} + \sqrt{(\mathbf{hr'}-)} = 100.59$ per cent; $\sqrt{(\mathbf{rh''}-)} + \sqrt{(\mathbf{hr''}-)} = 100.11$ per cent.

2. Parents who are **rh′** negative cannot have **hr′**-negative children, and **hr′**-negative parents cannot have **rh′**-negative children.
3. Similarly, **rh″**-negative parents cannot have **hr″**-negative children, and **hr″**-negative parents cannot have **rh″**-negative children.

Only rare exceptions to these rules have been encountered, all presumably accounted for by illegitimacy. Significantly, when mother-child combinations are studied, as in material from medicolegal cases of disputed paternity, no contradictions to the second law are found. One recent exception to this statement was encountered by Levine,[143] and evidence has been obtained that this baby had been interchanged with another at a foundling home.

Population Studies. Population studies in which Hr tests as well as Rh tests have been carried out yield data from which gene frequencies can be readily calculated. Formulae for these gene frequencies are derived, which complement those based on data for the 8 Rh types (cf. formula (31)–(36)). For example, phenotype $\mathrm{Rh_1Rh_1}$ includes genotypes R^1R^1 and R^1r', while phenotype rh′rh′ corresponds to genotype $r'r'$.

$$\text{Therefore, } R^1 + r' = \sqrt{\mathrm{Rh_1Rh_1} + \mathrm{rh'rh'}}$$

$$\text{Similarly, } R^2 + r'' = \sqrt{\mathrm{Rh_2Rh_2} + \mathrm{rh''rh''}}$$

Since phenotypes rh′rh′ and rh″rh″ are both extremely rare, the following approximations may be used.

$$R^1 = \sqrt{\mathrm{Rh_1Rh_1}} - r' = \sqrt{\mathrm{Rh_1Rh_1}} - \sqrt{\mathrm{rh'} + \mathrm{rh}} + \sqrt{\mathrm{rh}} \quad (38)$$

$$\text{similarly, } R^2 = \sqrt{\mathrm{Rh_2Rh_2}} - \sqrt{\mathrm{rh''} + \mathrm{rh}} + \sqrt{\mathrm{rh}} \quad (39)$$

The frequency of gene R^z can be estimated from that of phenotype $\mathrm{Rh_zRh_1}$, which includes genotypes R^zR^1, R^zr' and R^1r^y. If we disregard the latter two genotypes, which are rare in comparison to the first, it can be seen that the frequency of gene R^z is derived by dividing the frequency of phenotype $\mathrm{Rh_zRh_1}$ by twice the frequency of gene R^1, as given by the formula (38) above. As far as the frequency of the gene r^y is concerned, this is so low that a similar method of calculation cannot be used. However, the frequency of gene r^y can be estimated from the fact that in family studies, among Caucasoids, approximately half the parents of phenotype rh′rh″ prove to be genotype $r'r''$ and about half are genotype r^yr.

The rarest genotype is r^yr^y, corresponding to phenotype $\mathrm{rh_yrh_y}$, with an estimated frequency of about 1 in 100 million among Caucasoids. McGee et al.[144] have recently encountered such an individual and, as might be expected, he was derived from a consanguineous marriage.

When the gene frequencies are estimated with the aid of the formulae, their sum is found to be close to 100 per cent, as expected.

In view of the reciprocal relationship between **rh′** and **hr′**, and between **rh″** and **hr″** the following should hold.

$$\sqrt{\mathbf{rh'}\text{ neg}} + \sqrt{\mathbf{hr'}\text{ neg}} = 100 \text{ per cent} \qquad (40)$$

$$\text{and } \sqrt{\mathbf{rh''}\text{ neg}} + \sqrt{\mathbf{hr''}\text{ neg}} = 100 \text{ per cent} \qquad (41)$$

In table 33, Race's data on the distribution of the Rh-Hr types in London have been subjected to statistical analysis, and it can be seen that they satisfy the requirements of the genetic theory.

Serology. Anti-**hr′** serums are generally obtained from type Rh_1Rh_1 individuals who have been sensitized by pregnancy or by transfusion, and similarly, anti-**hr″** serums are obtained from type Rh_2Rh_2 isosensitized patients. Attempts to produce anti-**hr′** and anti-**hr″** serums in volunteer donors by deliberate injections of type rh blood have only occasionally been successful, which corresponds with the clinical observation that cases of **hr′** and **hr″** sensitization are far less common than cases of $\mathbf{Rh_0}$ sensitization. A type Rh_1Rh_1 individual sensitized to type Rh_1rh (or type rh) blood will produce pure anti-**hr′**, but if he is sensitized to type Rh_1Rh_2 blood he may produce the two antibodies anti-**hr′** and anti-**rh″**. (When an individual of the rare type Rh_ZRh_1 is isosensitized his serum will necessarily contain only anti-**hr′**.)[145] Polyvalent anti-**hr′rh″** serum is difficult to differentiate from pure anti-**hr′**, unless the two antibodies in the former have unequal titers or react by different methods of testing. With the aid of blood of the rare type Rh_ZRh_1, however, the distinction can easily be made since such blood is **hr′** negative but **rh″** positive. Obviously failure to recognize the presence of anti-**rh″** in an anti-**hr′** reagent can lead to the incorrect typing of type Rh_ZRh_1 blood as Rh_ZRh_0. Similarly, many anti-**hr″** reagents also contain anti-**rh′**, and here blood of the rare type Rh_ZRh_2 is valuable for eliminating this possible source of error.

An important peculiarity of anti-**hr′** and anti-**hr″** serums is that they give a much more pronounced gene dosage effect than Rh antiserums. As has already been pointed out, failure to take this into account has given rise to errors in Rh-Hr typing.

Additional Rh-Hr Factors

Factor rh^w. Callender and Race[146] detected the antibody for this blood factor in serum of a woman who had lupus erythematosis and who had been immunized by numerous transfusions. She belonged to type Rh_1Rh_1 and became sensitized to factors **hr′** and **N**, and to three previously undescribed blood factors, of which **rh**^{w} was one. The superscript "w" was chosen because the name of the patient was Willis. The incidence of the new factor is relatively low, only 3 to 7 per cent among Caucasoids. A striking feature of the factor is that all persons possessing the factor also possess the factor **rh′**. (Generally, however, blood having factor **rh**^{w} reacts more weakly with anti-**rh′** serum than blood lacking factor **rh**^{w}.) The existence of this factor therefore subdivides type Rh_1, for example, into two types, one with and the other without the **rh**^{w} factor. These findings make it necessary to increase the number of Rh allelic genes, since in place of gene R^1 there must be substituted two genes R^{1w} and gene R^1 "proper". Similarly, evidence has been obtained for the existence of two kinds of r' gene, namely r'^w and r' "proper".[147] It is possible that genes R^{zw} and r^{yw} also exist, but if they do, they must be extremely rare. Family studies on the heredity of the **rh**^{w} factor support the genetic theory.[148, 149]

Factor hr. Rosenfield et al.[150, 151] encountered an antiserum whose reactions at first seemed to correspond to those expected of anti-$\mathbf{Hr_0}$. Family and population studies showed that the new serum, designated anti-*f* by the discoverers, but to which the designation anti-**hr** has been assigned by Wiener,[151] reacted with blood cells from all individuals carrying genes r and R^0, and did not react with blood from persons lacking these genes.

In order to crystallize the relationships among the agglutinogens determined by the blood factors described up to this point, namely, $\mathbf{Rh_0}$, **rh′**, **rh**^{w}, **rh″**, **hr′**, **hr″**, and **hr**, the Rh-Hr allelic genes and their corresponding agglutinogens identifiable with the aid of antiserums for these seven blood factors are summarized in table 34. As can be seen, it is necessary to postulate a minimum of ten Rh allelic genes. As has already been pointed out, the discovery of the **hr** factor requires no extension of the series of allelic genes, since it merely defines

TABLE 34

THE Rh-Hr ALLELIC GENES IDENTIFIABLE WITH ANTISERUMS FOR FACTORS **Rh**$_0$, **rh′**, **rh**^{w}, **rh″**, **hr′**, **hr″** and **hr**.

Genes	Frequency among Caucasoids (per cent)	Corresponding Agglutinogens	Blood Factors Present
r	38.	rh	**hr′**, **hr″**, **hr**
r'	0.6	rh′	**rh′**, **hr″**
r'^w	0.005	rh′^{w}	**rh′**, **rh**^{w}, **hr″**
r''	0.5	rh″	**rh″**, **hr′**
r^y	0.01	rh^y	**rh′**, **rh″**
R^0	2.7	Rh$_0$	**Rh**$_0$, **hr′**, **hr″**, **hr**
R^1	41.	Rh$_1$	**Rh**$_0$, **rh′**, **hr″**
R^{1w}	2.	Rh$_1^w$	**Rh**$_0$, **rh′**, **rh**^{w}, **hr″**
R^2	15.	Rh$_2$	**Rh**$_0$, **rh″**, **hr′**
R^z	0.2	Rh$_z$	**Rh**$_0$, **rh′**, **rh″**

Omitted from the table are the genes r^{yw} and R^{zw}, both presumably very rare, and of which only a single family with r^{yw} has been found to date.

an additional serological attribute which is shared by the two agglutinogens rh and Rh$_0$. These two agglutinogens could formerly be identified without the aid of serum anti-**hr**. However, the antiserum does help in distinguishing between individuals of genotypes $r'r''$ and r^yr by serological means alone, because the former are **hr** negative while the latter are **hr** positive. Similarly, in persons of type Rh$_1$Rh$_2$, the use of this antiserum helps in making the distinction between those who do and those who do not carry the gene R^z, and this not infrequently is of value in cases of disputed parentage.

The number of genotypes possible with n alleles can be shown to be equal to $\frac{1}{2}n(n + 1)$, so that the 10 Rh-Hr alleles give rise to 55 possible genotypes. When all of the seven antiserums that have been described thus far are available, as many as 28 phenotypes can be distinguished corresponding to these 55 genotypes. For medicolegal work the use of antiserums for factors **Rh**$_0$, **rh′**, **rh″**, and **hr′** is standard practice, while the less readily available antiserums anti-**hr″**, anti-**rh**^{w} and anti-**hr** are used only in a limited number of laboratories. Table 35 is a systematized tabulation of the phenotypes which can be identified according to the reagents used, together with their corresponding genotypes.

Factor hr^v. In 1956, DeNatale et al.[153] encountered an antiserum for still another factor of the Rh-Hr system. This factor proved to

TABLE 35

THE Rh-Hr PHENOTYPES AND GENOTYPES

International Nomenclature

2 Rh Phenotypes			12 Rh Phenotypes					28 Rh-Hr Phenotypes					55 Genotypes*
					Reaction with					Reaction with			
Designations	Approximate frequencies in N. Y. C. whites (%)	Reaction with anti-**Rh**$_0$ (or anti-rhesus)	Designation†	Approximate frequencies in N. Y. C. whites (%)§	Anti-**rh**′	Anti-**rh**″	Anti-**rh**^{w}	Designation	Approximate frequencies in N. Y. C. whites (%)§	Anti-**hr**′	Anti-**hr**″	Anti-**hr**	
Rh negative	15	−	rh	14.4	−	−	−	rh	14.4	+	+	+	*rr*
			rh′	0.46‡	+	−	−	rh′rh	0.46	+	+	+	$r'r$
								rh′rh′	.0036	−	+	−	$r'r'$
			rh'^{w}	.004	+	−	+	$rh'^{w}rh$	.004	+	+	+	$r'^{w}r$
								$rh'^{w}rh'$	.00006	−	+	−	$r'^{w}r'$ or $r'^{w}r'^{w}$
			rh″	0.38	−	+	−	rh″rh	0.38	+	+	+	$r''r$
								rh″rh″	.0025	+	−	−	$r''r''$
			rh_y	.01	+	+	−	rh′rh″	.006	+	+	−	$r'r''$
								rh_yrh	.008	+	+	+	$r^{y}r$
								rh_yrh'	.0001	−	+	−	$r^{y}r'$
								rh_yrh''	.0001	+	−	−	$r^{y}r''$
								rh_yrh_y	.000001	−	−	−	$r^{y}r^{y}$
			rh_y^{w}	.00005	+	+	+	$rh'^{w}rh''$	.00005	+	+	−	$r'^{w}r''$
								$rh_y^{w}rh'$	.000001	−	+	−	$r'^{w}r^{y}$

Rh positive	85	+	Rh_0	2.1	−	−	−	Rh_0	2.1	+	+	+	R^0R^0 or R^0r
			Rh_1	50.7	+	−	−	Rh_1rh	33.4	+	+	+	R^1r, R^1R^0, or R^0r'
								Rh_1Rh_1	17.3	−	+	−	R^1R^1 or R^1r'
			Rh_1^w	3.3	+	−	+	Rh_1^wrh	1.6	+	+	+	$R^{1w}r$, $R^{1w}R^0$, or $R^0r'^w$
								$Rh_1^wRh_1$	1.7	−	+	−	$R^{1w}R^1$, $R^1r'^w$, $R^{1w}r'$, $R^{1w}R^{1w}$, or $R^{1w}r'^w$
			Rh_2	14.6	−	+	−	Rh_2rh	12.2	+	+	+	R^2r, R^2R^0, or R^0r''
								Rh_2Rh_2	2.4	+	−	−	R^2R^2 or R^2r''
			Rh_Z	13.4	+	+	−	Rh_1Rh_2	12.9	+	+	−	R^1R^2, R^1r'', or R^2r'
								Rh_Zrh	0.2	+	+	+	R^zr, R^zR^0, or R^0r^y
								Rh_ZRh_1	0.2	−	+	−	R^zR^1, R^zr', or R^1r^y
								Rh_ZRh_2	.07	+	−	−	R^zR^2, R^zr'', or R^2r^y
								Rh_ZRh_Z	.0004	−	−	−	R^zR^z or R^zr^y
			Rh_Z^w	0.6	+	+	+	$Rh_1^wRh_2$	0.6	+	+	−	$R^{1w}R^2$, $R^{1w}r''$ or $R^2r'^w$
								$Rh_Z^wRh_1$	.008	−	+	−	$R^{1w}R^z$, $R^{1w}r^y$, or $R^zr'^w$

* This table does not include genes R^{zw} and r^{yw}, which appear to be very rare.

† In this table Rh_1 is used as a short designation for Rh_0'; Rh_2 is short for Rh_0''; rh_y is short for rh′ ″; and Rh_Z is short for Rh'_0''.

‡ The reduction in the frequency of type rh′ as compared with that given in earlier charts can be attributed to recognition of bloods of type $\mathfrak{R}h_1$ (containing Rh_0 variant) which are now included in type Rh_1 instead of rh′. The agglutinogens $\mathfrak{R}h_0$, $\mathfrak{R}h_1$, and $\mathfrak{R}h_2$, and their corresponding genes $\mathfrak{R}^0$, $\mathfrak{R}^1$, and $\mathfrak{R}^2$, are not given here, because this would serve unnecessarily to complicate the chart, by increasing the number of possible genotypes to 91. Also, no attempt is made to include certain rare exceptional bloods, such as those lacking both factors **rh′** and **hr′**, and/or lacking both **rh″** and **hr″**, etc.

§ Based on the estimated gene frequencies, $r = 0.38$, $r' = .006$, $r'' = .005$, $r^y = .0001$, $r'^w = .00005$, $R^0 = .027$, $R^1 = 0.41$, $R^2 = 0.15$, $R^z = .002$, and $R^{1w} = .02$.

have an incidence of approximately 40 per cent among Africans, 27 per cent in New York City Negroids, and only about 0.5 per cent among Caucasoids. It was noted that all individuals possessing the factor were also carriers of the genes r or R^0 or both, so that whenever the new factor, named V, is present, factor **hr** was also present. Thus, this factor bears the same relationship to factor **hr** that factor $\mathbf{rh}^{w}$ bears to factor **rh′**. For this reason it is proposed to call this factor $\mathbf{hr}^{v}$. Evidently in Negroids there are at least two varieties of gene r and gene R^0, namely, gene r^v and gene r "proper", and gene R^{0v} and gene R^0 "proper", respectively. The use of anti-$\mathbf{hr}^{v}$ serum increases the number of phenotypes that can be distinguished, and appropriate names for these can readily be devised.

Complexities of the Rh_0 Factor and Associated Factors

Old blood suspensions generally give weaker agglutination reactions than fresh blood suspensions. Even when fresh blood exclusively is tested, variations in intensity of the reactions are encountered, dependent on the individual's blood type. For example, as already pointed out, blood cells from homozygous individuals not infrequently give more intense reactions than blood from heterozygous individuals, the so-called gene dose effect. In addition, group A_2B cells react more weakly than A_2 cells with anti-**A** serums, and type Rh_1 cells react more weakly than type rh′ cells with anti-**rh′** serums. This suggests that the A-B-O and Rh-Hr agglutinogens are each formed from a separate substrate, and the presence of other structures in the agglutinogen molecule may interfere with the agglutinability of the cells, e.g., the presence of factor $\mathbf{Rh}_0$ in type Rh_1 cells interferes with the **rh′**/anti-**rh′** reaction, and the presence of $\mathbf{rh}^{w}$ still further reduces the agglutinability of the cells by anti-**rh′**. Aside from such phenomena, agglutinability may be affected by minor variations in the structure of the agglutinogen molecule itself causing corresponding minor changes in the blood factors, such as in agglutinogens A_1, A_2, and A_3, agglutinogens N and N_2, etc. Similar variations in the Rh-Hr agglutinogens, especially those involving factor $\mathbf{Rh}_0$ will now be discussed.

In 1944, Wiener[154] showed that there were certain blood specimens which gave weak but distinct reactions with certain of the Rh antiserums. These blood specimens appear to occupy a position intermediate between negatively-reacting bloods and those which gave sharp positive reactions. Family studies indicated that this characteristic is hereditary and because of this and their serological behavior these types were designated as "intermediate" Rh types and the genes called "intermediate" genes. Of the various intermediate Rh-Hr variants the **Rh**$_0$ variants designated $\mathfrak{R}h_0$ are the most important. As will be seen, recent investigations confirm the impression that the **Rh**$_0$ factors hold a special central position in the Rh-Hr system.

Factor $\mathfrak{R}h_0$. The so-called factor $\mathfrak{R}\mathbf{h}_0$ actually comprises a group of factors giving reactions of intermediate intensity with anti-**Rh**$_0$ sera, and the symbol $\mathfrak{R}h_0$ has been used for the entire group of variants. Blood specimens which react typically with most anti-**Rh**$_0$ serums but give slightly weaker reactions with a number are known as high-grade variants, while those on the other end of the scale reacting consistently weakly with all anti-**Rh**$_0$ serums are low-grade variants. Most of the variants give similar reactions which are characterized as follows. In saline media, they fail to react with anti-**Rh**$_0$ serums, even with those containing potent anti-**Rh**$_0$ agglutinins. They frequently, but not always, react with univalent **Rh**$_0$ antiserums by the conglutination method. They almost always react with such serums by the ficin technique, and characteristically react with such serums by the anti-globulin technique, although the avidity and titer of the reactions are generally lower than for typical **Rh**$_0$-positive bloods. Interestingly, the bloods of chimpanzees give reactions resembling those of high grade **Rh**$_0$ variants.

In 1945, Wiener, Unger and Sonn[155] pointed out that **Rh**$_0$ variants are relatively rare among Caucasoids and occur more frequently among Negroids. Moreover, it has been found that the factor $\mathfrak{R}\mathbf{h}_0$ like the factor **Rh**$_0$ may occur alone or together with **rh′**, **rh″**, or both. To account for these findings, the existence of at least three additional allelic genes has been postulated, namely, $\mathfrak{R}^0$, $\mathfrak{R}^1$, and $\mathfrak{R}^2$, corresponding to the agglutinogens $\mathfrak{R}h_0$, $\mathfrak{R}h_1$, and $\mathfrak{R}h_2$, respectively.

Further studies revealed that certain bloods which previously had been designated as type rh′ or type rh″ were actually type $\mathfrak{R}h_1$ or $\mathfrak{R}h_2$, respectively. An extensive study on this question was carried out by Rosenfield et al.[156] on a series of white blood donors in New York City. The blood specimens were first tested with saline agglutinating anti-**Rh**$_0$, anti-**rh′**, and anti-**rh″** serums, and then all apparently **Rh**$_0$-negative bloods were tested for **Rh**$_0$ variants using the anti-globulin technique. It was found that of 161 seeming type rh′rh, as many as 72 or 44.7 per cent were actually type $\mathfrak{R}h_1rh$; of 92 seeming type rh″rh, as many as 19 or 20.7 per cent proved to be type $\mathfrak{R}h_2rh$; while of 628 seeming type rh, only 3 or 0.5 per cent were actually type $\mathfrak{R}h_0$. These results indicate that among Caucasoids, of the three postulated additional rare alleles, $\mathfrak{R}^1$ is the most frequent, $\mathfrak{R}^2$ is next in frequency, while $\mathfrak{R}^0$ is by far the rarest.

These findings argue against the idea that the existence of the $\mathbf{Rh_0}$ variants is due to one or more modifying genes, for unless such genes were completely linked to the Rh-Hr genes, the distribution of the $\mathbf{Rh_0}$ variants would be expected to be largely independent of the Rh-Hr types. A disturbing finding is the not infrequent occurrence of individuals of type Rh_1Rh_1 whose blood cells give weak reactions with anti-$\mathbf{Rh_0}$ serums, resembling those of high grade variants. By family studies Ceppelini[157] has shown that these persons are always of genotype R^1r', and that in other type Rh_1 individuals in the same family, such as those of genotype R^1r, the bloods give reactions of typical $\mathbf{Rh_0}$-positives. Ceppelini therefore believes that the r' gene in these families has a special modifying action, which is responsible for the weak reactions obtained with $\mathbf{Rh_0}$ antiserums with the red cells of such individuals. Levine et al.[158] have recently reported a large pedigree confirming Ceppelini's hypothesis, and similar observations have been made by Wiener.

The practical clinical importance of the $\mathbf{Rh_0}$ variants is that transfusion of such blood to an $\mathbf{Rh_0}$-negative individual may give rise to sensitization to the $\mathbf{Rh_0}$ factor. As will be discussed later on in greater detail, a few $\mathbf{Rh_0}$-positive individuals have become sensitized to a factor related to the $\mathbf{Rh_0}$ factor.

Nomenclature. The difference between factors $\mathbf{Rh_0}$ and $\mathfrak{R}\mathbf{h_0}$ is not on the same plane as the difference among $\mathbf{Rh_0}$, $\mathbf{rh'}$ and $\mathbf{rh''}$, although both are of a qualitative nature. Thus, although the reactions of factor $\mathbf{Rh_0}$ include all of the reactions of factor $\mathfrak{R}\mathbf{h_0}$ the reverse is not true, since $\mathfrak{R}\mathbf{h_0}$ shares only some of the reactions of $\mathbf{Rh_0}$. Moreover, the difference between $\mathbf{Rh_0}$ and $\mathfrak{R}\mathbf{h_0}$ is not on the same plane as that between $\mathbf{rh'}$ and $\mathbf{rh^w}$, because while serum for $\mathbf{rh^w}$ does not react with $\mathbf{rh'}$, serum for $\mathfrak{R}\mathbf{h_0}$ always reacts with $\mathbf{Rh_0}$. Type rh individuals immunized with type $\mathfrak{R}h_0$ blood produce antiserums which, like ordinary anti-$\mathbf{Rh_0}$ serums react more strongly with blood having the $\mathbf{Rh_0}$ factor than those containing the $\mathfrak{R}\mathbf{h_0}$ factor. Thus, in the case of factors $\mathbf{Rh_0}$ and $\mathfrak{R}\mathbf{h_0}$, the distinctions made are not those of sharply differentiated blood factors, but among members of a graded series of related factors. To indicate this it seems more appropriate to change the type face of the capital R to Germanic capital $\mathfrak{R}$ than to coin a new symbol. As will be pointed out, there are also variants of the $\mathbf{Rh_0}$ factor which react more strongly with anti-$\mathbf{Rh_0}$ serum than does ordinary $\mathbf{Rh_0}$-positive blood. To indicate this the symbol of these $\mathbf{Rh_0}$ variants is changed by using a capital R with a bar over it: $\bar{\mathrm{R}}$.

Variants of the factors **rh′** and **rh″** also have been encountered, and in the past the custom has been to enclose the prime sign and the double-prime sign in parentheses to indicate such variants. Like bloods with variants of factor $\mathbf{Rh_0}$, bloods with variants of factors **rh′** and **rh″** are rare.

Factor $\bar{\mathrm{R}}\mathrm{h_0}$. In 1950 Race, Sanger and Selwyn[159] found in the serum of the mother of a baby who died of erythroblastosis fetalis an antibody which clumped 1,400 consecutive blood specimens, but failed to clump her own cells. The mother's red cells proved to be unique in that they reacted with anti-$\mathbf{Rh_0}$ serum but appeared to lack both factors of the pairs **rh′-hr′** and **rh″-hr″**. To account for this observation, these investigators postulated that a deletion had occurred in the Rh "chromosomes", and asserted that the genotype of the mother was *–D–/–D–* (cf. page 97).

The following interpretation appears more consistent with established serological and genetical principles. As Race et al. noticed, in addition to the peculiarities mentioned above, the red cells were agglutinated in saline media with anti-$\mathbf{Rh_0}$ blocking serum, which failed to clump ordinary $\mathbf{Rh_0}$-positive cells under identical conditions. Thus, this individual had an agglutinogen similar to agglutinogen $\mathrm{Rh_0}$, but differing in its higher reactivity with anti-$\mathbf{Rh_0}$ serum.[160] It is this property which we consider the fundamental characteristic of this agglutinogen, due to a modification in its chemical structure which increases the reactivity with anti-$\mathbf{Rh_0}$ serums, and simultaneously crowds out the site of reactivity with all anti-Hr serums. This new type of antigen we therefore propose to designate "super"-$\mathbf{Rh_0}$ and assign the symbol "bar" $\mathbf{Rh_0}$ (or $\bar{\mathbf{R}}\mathbf{h_0}$) to its blood factor. Corresponding to agglutinogen $\bar{\mathrm{R}}\mathrm{h_0}$ the existence of another allelic gene $\bar{R}^0$ is postulated.

The existence of gene $\bar{R}^0$ can give rise to contradictions to the rule that an **rh**′-negative parent cannot have an **hr**′-negative child, and an **rh**″-negative parent cannot have an **hr**″-negative child, and this must be taken into account in the medicolegal applications of the tests. For example, a mother whose genotype is $R^1\bar{R}^0$, and who is therefore **hr**′ negative, could have a child who is **rh**′-negative. Fortunately, the gene $\bar{R}^0$ betrays its presence, even in its heterozygous form, by the agglutination of the individual's red cells with univalent anti-$\mathbf{Rh_0}$ serums in saline media. In practice such cases will rarely occur because of the extreme rarity of gene $\bar{R}^0$, but the medicolegal expert must be aware of this possibility if serious mistakes are to be avoided.

The gene $\bar{R}^0$ and its corresponding agglutinogen $\bar{\text{R}}\text{h}_0$ will probably prove to be the prototype of a whole series of such Rh-Hr genes and corresponding agglutinogens. Thus, corresponding to the variants $\mathfrak{R}\text{h}_0$, $\mathfrak{R}\text{h}_1$, $\mathfrak{R}\text{h}_2$, etc. one may anticipate the existence in addition to $\bar{\text{R}}\text{h}_0$ of $\bar{\text{R}}\text{h}_1$, and $\bar{\text{R}}\text{h}_2$, the latter being characterized by their "super" $\mathbf{Rh_0}$ factors. In fact, Gunson and Donahue[161] have found an individual of type $\bar{\text{R}}\text{h}^{\text{w}}$, who by family studies was shown to be homozygous for a new allelic gene which may be designated as $\bar{R}^w$. His red cells lacked the four blood factors **rh**′, **hr**′, **rh**″ and **hr**″.

The feature of both blood specimens, $\bar{\text{R}}\text{h}_0$ and $\bar{\text{R}}\text{h}^{\text{w}}$, which attracted the attention of the discoverers was the simultaneous absence of both factors of the pairs **rh**′-**hr**′ and/or **rh**″-**hr**″. According to the interpretation proposed here, this is to be expected in any individual homozygous for a "super"-R^0 gene. Such individuals are extremely rare, and it is therefore not surprising that in both cases they were the products of consanguineous marriages. To detect the more common heterozygotes, advantage should be taken of what we regard as the more fundamental characteristic of these agglutinogens, namely, their high agglutinability with anti-$\mathbf{Rh_0}$ serums. Ordinary Rh-Hr typing would not detect these carriers of a super R^0 gene since the other gene of the pair determining the genotype would provide reactions for each of the factor pairs **rh**′-**hr**′ and **rh**″-**hr**″.

Theoretical Serological Considerations. Up to the present it has been impossible to demonstrate conclusively the existence of anti-$\mathbf{Hr_0}$, although theoretically there seems to be no reason why a factor bearing the same reciprocal relationship to factor $\mathbf{Rh_0}$ that factor **hr**′ has to factor **rh**′ and factor **hr**″ has to factor **rh**″ should not exist. A possible explanation for this may be found in the following analysis, based on the known facts concerning factor **Rh**$_0$, and its variants $\mathfrak{R}$h$_0$ and $\bar{\text{R}}$h$_0$.

It has already been pointed out how the presence of one structure of factor in the Rh-Hr agglutinogen molecule may interfere with the development of other Rh-Hr factors, as judged from the serological reactions. Similarly, it might be anticipated that the more prominent the structure responsible for the reactions which are obtained with anti-**Rh**$_0$ serums, the less prominent would be the hypothetical **Hr**$_0$ factors. On this basis, the following chart may be drawn up.

Agglutinogens	Known **Rh**$_0$ factors	Hypothetical **Hr**$_0$ factors
rh, rh′, rh″, etc.	none	$\bar{\mathbf{H}}$**r**$_0$
$\mathfrak{R}$h$_0$, $\mathfrak{R}$h$_1$, $\mathfrak{R}$h$_2$, etc.	$\mathfrak{R}$**h**$_0$	**Hr**$_0$
Rh$_0$, Rh$_1$, Rh$_1^w$, Rh$_2$, etc.	**Rh**$_0$	$\mathfrak{H}$**r**$_0$
$\bar{\text{R}}$h$_0$, $\bar{\text{R}}$h$_1^w$	$\bar{\mathbf{R}}$**h**$_0$	none

It will be seen that the variations of the hypothetical **Hr**$_0$ factors are indicated by changing the type face of the capital H in the symbol in conformity with the symbols used for the various **Rh**$_0$ factors. As is well known, **Rh**$_0$ antibodies are readily produced by persons lacking the **Rh**$_0$ factor. Until recently no case was known of **Rh**$_0$ antibodies being produced by individuals possessing the **Rh**$_0$ factor. However, instances of the production of anti-**Rh**$_0$ by persons having the rare $\mathfrak{R}$h$_0$ variants are now well documented, though such cases are quite rare. It is postulated that the difficulty encountered in finding anti-**Hr**$_0$ up until now has been due to the fact that it has been searched for in individuals homozygous for the ordinary **Rh**$_0$ factor, such as persons of type Rh$_1$Rh$_2$. As shown by the chart, however, if the concept proposed is correct, such persons have a variant of the **Hr**$_0$ factor and therefore would not be apt to respond readily to injections of **Hr**$_0$-positive blood. The only individuals who might be expected to form this antibody readily would be those homozygous for a gene determining an agglutinogen having the $\bar{\mathbf{R}}$**h**$_0$ factor since only such individuals would be truly **Hr**$_0$ negative according to the hypothesis. In fact, Buchanan and McIntyre[162] found in the serum of his subject an antibody which agglutinated all bloods tested except his own. A similar antibody was found by Gunson and Donahue in their $\bar{R}^w\bar{R}^w$ individual, and significantly the bloods of the $\bar{R}^0\bar{R}^0$ person and the $\bar{R}^w\bar{R}^w$ person proved to be compatible. Thus, it is likely according to this hypothesis that the antibodies in both instances were of the specificity anti-$\mathbf{Hr}_0$. Since the anti-$\mathbf{Hr}_0$ described here reacts with the blood of all persons except those homozygous for the rare super R^0 gene, it is quite different from the anti-*d* postulated by Fisher and the hypothetical anti-$\mathbf{Hr}_0$ of table 30.

Factors Rh^A, Rh^B, Rh^C, ... Wiener and Geiger[163, 164] studied a case of erythroblastosis fetalis in which the mother was $\mathbf{Rh}_0$ positive, and yet her serum contained an antibody giving all of the reactions expected of anti-$\mathbf{Rh}_0$ except that it failed to clump her own cells. The reactions of the mother's cells indicated that they did not merely contain an $\mathbf{Rh}_0$ variant, as in the case reported by Argall et al.[165] Moreover, serum from Argall's case clumped the cells of Wiener and Geiger's patient and conversely. Rosenfield et al.[166] investigated another case, similar to that of Wiener and Geiger, but again cross matching of the blood specimens proved the antibodies involved to be different. To explain these observations. Wiener suggested that the blood of $\mathbf{Rh}_0$-positive individuals have associated with the $\mathbf{Rh}_0$ factor, a series of factors $\mathbf{Rh}^A$, $\mathbf{Rh}^B$, etc. In certain rare individuals, one or more of the associated factors may be missing and such individuals can be sensitized to the missing blood factor. Since "normal" $\mathbf{Rh}_0$-positive blood has all of the factors $\mathbf{Rh}^A$, $\mathbf{Rh}^B$, $\mathbf{Rh}^C$, etc., anti-$\mathbf{Rh}^A$ serum, for example, is indistinguishable from anti-$\mathbf{Rh}_0$ serum except that it fails to agglutinate blood cells of rare individuals lacking the factor $\mathbf{Rh}^A$. If this hypothesis is correct, it would be expected that ordinary anti-$\mathbf{Rh}_0$ serum could contain a mixture of the antibodies anti-$\mathbf{Rh}^A$, anti-$\mathbf{Rh}^B$, anti-$\mathbf{Rh}^C$ . . . etc. Indeed it was found that when anti-$\mathbf{Rh}_0$ serum was absorbed with cells of type Rh_0^a a fraction of the antibodies remained which gave reactions corresponding to anti-$\mathbf{Rh}^A$.

To designate those rare Rh-positive bloods which lack one of the associated factors $\mathbf{Rh}^A$, $\mathbf{Rh}^B$, . . . an appropriate symbol indicating which factor is absent is used. For example, a type Rh_1^a individual, like the mother of the erythroblastotic baby studied by Wiener and Geiger, has factor $\mathbf{Rh}_0$ and all its associated factors except $\mathbf{Rh}^A$. Studies[163] on a large series of Caucasoids showed that Rh-positive bloods lacking factor $\mathbf{Rh}^A$ are extremely rare, but among Negroids as many as one out of 200 Rh-positive persons were found to lack factor $\mathbf{Rh}^A$. If, as seems reasonable, the atypical blood types described are prototypes of a series of blood types yet to be discovered, it becomes necessary to postulate the existence of additional rare alleles in the Rh-Hr series of allelic genes such as R^{0a}, R^{1a} . . . R^{0b}, R^{1b} . . . etc. Fortunately, these genes, if they exist at all, are extremely rare, so that these new discoveries will have little effect on routine Rh-Hr typing.

Additional Complexities of the Rh-Hr Types

Just as the $\mathbf{Rh_0}$ factor has its variants, so is there evidence of the existence of rare variants of the other Rh-Hr factors. Suitable terminology for designating variants of **rh′** and **rh″** has already been indicated, and similar symbols can be devised for the variants of the other Rh-Hr factors. However, knowledge concerning these rare variants is limited, and it is best not to let the symbols get ahead of the facts, or changes in terminology may become necessary later on and lead to confusion.

An antibody has been encountered for the rare blood factor, which like $\mathbf{rh^w}$, occurs only in association with **rh′**. This antibody [167] may be designated anti-$\mathbf{rh^x}$ so that at least three sharply defined types of Rh_1 agglutinogen exist, namely, Rh_1^w, Rh_1^x, and Rh_1 "proper", the latter being, of course, the most common. Moreover, a rare factor has been discovered which occurs only in association with factor **rh″**. This antibody[168] may be designated anti-$\mathbf{rh^{w2}}$ to distinguish it from the anti-$\mathbf{rh^w}$ serum for the factor associated with **rh′**.

A final complexity may be mentioned and that is capacity of $\mathbf{Rh_0}$-negative individuals sensitized with $\mathbf{Rh_0}$-positive blood, lacking the **rh′** factor, to produce antibodies which react with type rh′ blood, which lacks the factor $\mathbf{Rh_0}$. Waller et al.[169] obtained one such serum from an immunized donor which reacted with almost equal titer on type Rh_0 and type rh′ blood and could be absorbed completely with either. Thus, factors **rh′** and $\mathbf{Rh_0}$ appear to have at least one factor in common, and this has been designated $\mathbf{rh^G}$ by Allen. By testing anti-$\mathbf{rh^G}$ serum against the red cells of individuals of type rh, Allen encountered one rare blood which reacted with the serum, and this was therefore designated as type rh^G. Wiener has pointed out the analogy between type rh^G blood and group A_0 blood, since the latter in ordinary tests reacts like group O blood, but is agglutinated by serum from most group O individuals which contain anti-**C**.

In conclusion, it is evident that in addition to the 10 "standard"

Rh genes r, r', r'^{w}, r'', r^{y}, R^{0}, R^{1}, R^{1w}, R^{2}, and R^{z}, it is now necessary to postulate additional rare alleles r^{v}, r^{G}, $R^{0\mathrm{v}}$, R^{0a}, R^{0b} . . . $\mathfrak{R}^{0}$, $\bar{R}^{0}$, R^{1a}, R^{1b}, . . . R^{1x}, R^{2w_2}, R^{2a}, R^{2b}, . . . etc.

C-D-E Notations for the Rh-Hr Types

Historical Background. While in this review, Wiener's Rh-Hr nomenclature has been used exclusively, those who consult the literature will find that the majority of the articles use the Fisher-Race C-D-E notations, either exclusively or in combination with the Rh-Hr nomenclature. Therefore, it is necessary to explain the C-D-E notations and the concepts on which they are based.

Fisher's hypothesis as presented by Race[171] in 1946 is summarized in table 36. This table was explained by Race as follows:

"Eighteen months ago we reached this stage in agreement with, but independently of Wiener. Wiener lacked the very great help of *St* serum (*Ed. Note: St* serum is the anti-**hr**′ serum of Race and Taylor described on page 71). We brought our notation into line with Wiener's. In January 1944, Fisher examined table 37 and noticed that when a gene was positive with *St* it was negative with *rh*′. He supposed that the antigens disclosed by these two sera were due to alleles, and these he called *C* and *c*. To the antibodies he gave corresponding Greek letters, *St* becoming γ, and *rh*′, Γ. He considered that the antigen recognized by anti-Rh_0 which he called *D* had an allele *d*, and the antiserum for which was yet to be found. The antigen recognized by anti-*rh*″ he called *E* and the serum he called H. The antiserum for the hypothetical allele *e* was also to be found.

"In other words, there were three tightly linked loci on the chromosome responsible for *Rh*, making eight possible combinations in a chromosome (table 37) . . . . Briefly, Mourant found η, and the anti-*Hr* described by Waller and Levine* has the reaction predicted by Fisher for δ."

Shortly after the original report[172] describing Fisher's hypothesis, Cappell[173] suggested that instead of using Greek letters for the Rh-Hr antibodies the designations anti-*C*, anti-*D*, anti-*E*, anti-*c*, anti-*d*, and anti-*e* be used. This modification has been adopted by Fisher and Race.

Summarizing, Fisher postulated three pairs of contrasting genes,

* In a footnote, Race adds, "This apparent agreement has since been claimed by Levine to be due to a typographical error in his own paper."

TABLE 36

RACE'S SUMMARY OF FISHER'S HYPOTHESIS

Antisera	Genes	Antisera
Γ(Anti-*Rh′*)	*C* or *c*	γ (*St*)(*Hr*)
Δ(Anti-Rh_0)	*D* or *d*	δ } then to be found
H(Anti-*Rh″*)	*E* or *e*	η } then to be found

TABLE 37

FISHER'S SYNTHESIS
(Modified after R. R. Race)

Name of serum		{Original Modern	Anti-*Rh′* Anti-**rh′**	Anti-*Hr* Anti-**hr′**	Anti Rh_0 Anti -$\mathbf{Rh_0}$	Anti-*Rh″* Anti-**rh″**	Not yet found Anti-$\mathbf{Hr_0}$	Not yet found Anti-**hr″**
Antibody present		{Fisher Cappell	Γ Anti-*C*	γ Anti-*c*	Δ Anti-*D*	H Anti-*E*	δ Anti-*d*	η Anti-*e*
Genes								
Original name	Present name	Fisher						
Rh_z	R^z	*CDE*	(+)	(−)	(+)	(+)	(−)	(−)
Rh_1	R^1	*CDe*	+	−	+	−	(−)	(+)
Rh_y	r^y	*CdE*	+	−	−	+	(+)	(−)
Rh′	*r′*	*Cde*	+	−	−	−	(+)	(+)
Rh_2	R^2	*cDE*	−	+	+	+	(−)	(−)
Rh_0	R^0	*cDe*	−	+	+	−	(−)	(+)
Rh″	*r″*	*cdE*	−	+	−	+	(+)	(−)
rh	*r*	*cde*	−	+	−	−	(+)	(+)

which determine corresponding pairs of agglutinogens, *C-c*, *D-d*, and *E-e*, respectively. The correspondence between these symbols and those used by Wiener is shown in table 38. It will be seen that Fisher treats as a gene or agglutinogen what Wiener considers a blood factor, or only one of the multiple serological attributes of a given agglutinogen. No clear distinction among gene, blood factor, and agglutinogen is made in the Race-Fisher theory, and the three terms are used interchangeably. According to Fisher, the three gene pairs *C-c*, *D-e*, and *E-e* are assumed to be closely linked within the same chromosome. Thus, eight "chromosomes" are possible, corresponding to Wiener's eight "standard" allelic genes, as shown in table 39.

As has already been pointed out, the concept of linked genes had earlier been considered by Wiener and disproved and discarded (cf. page 63).

TABLE 38

CORRESPONDENCE BETWEEN FISHER'S AND WIENER'S SYMBOLS FOR ELEMENTARY ANTIGENS

Fisher's Genes or Agglutinogens	Wiener's Blood Factors
C	**rh′**
c	**hr′**
D	$\mathbf{Rh_0}$
d	$(\mathbf{Hr_0})$*
E	**rh″**
e	**hr″**

* Hypothetical. Wiener now uses this symbol in a dfferent sence (cf. page 89).

TABLE 39

COMPARISON OF THE FISHER-RACE LINKED GENE HYPOTHESIS AND WIENER'S MULTIPLE ALLELE THEORY

Fisher-Race			Wiener		
"Chromosomes"					
Fisher's symbols	"Short-hand" symbols*	Corresponding antigens	Genes	Corre-sponding aggluti-nogens	Blood factors
cde	r	*c*, *d*, and *e*	r	rh	**hr′** and **hr″**
Cde	R'	*C*, *d*, and *e*	r'	rh′	**rh′** and **hr″**
cdE	R''	*c*, *d*, and *E*	r''	rh″	**rh″** and **hr′**
CdE	R^y	*C*, *d*, and *E*	r^y	rh_y	**rh′** and **rh″**
cDe	R_0	*c*, *D*, and *e*	R^0	Rh_0	$\mathbf{Rh_0}$, **hr′** and **hr″**
CDe	R_1	*C*, *D*, and *e*	R_1	Rh_1	$\mathbf{Rh_0}$, **rh′** and **hr″**
cDE	R_2	*c*, *D*, and *E*	R^2	Rh_2	$\mathbf{Rh_0}$, **rh″** and **hr′**
CDE	R_z	*C*, *D*, and *E*	R_z	Rh_Z	$\mathbf{Rh_0}$, **rh′** and **rh″**

* For compactness, use is made by Fisher and Race of symbols which they designate "shorthand". As Race points out, these are based on Wiener's original symbols. However, the same Rh "shorthand" symbols are used by Fisher and Race for both genotypes and phenotypes.

Actually, there were three predictions implicit in the Fisher hypothesis, namely, that antibodies of specificity anti-*d* and anti-*e* would be found, and that crossing-over would be encountered among the components of the tightly linked gene complexes. Of these three predictions only one has been realized, namely, the discovery of anti-*e* (anti-**hr″**), as has been described elsewhere (cf. page 73). The various claims regarding the discovery of anti-*d* (anti-$\mathbf{Hr_0}$) could not be confirmed, and as Race and Sanger state these may have been examples of anti-*f* (anti-**hr**). In any event anti-*d* is not available. The anti-$\mathbf{Hr}_0$ serum described on page 89, is quite different from the anti-*d* serum conceived by Fisher.

Crossing Over. If crossing over could be shown to occur among the postulated pairs of linked genes, as proposed by Fisher, this would be important evidence in favor of his concept. In actual practice, however, this has never been observed.* But, if the genes are assumed to be closely linked, crossing over might occur too rarely to be detected by family studies. Fisher[175, 176] suggested, in support of his theory, that the rarity of crossing-over actually accounted for the low incidence of the rarer genetic combinations in the population. For example, in Caucasoids the less common "chromosomes" *cDe* (R^0), *Cde* (r'), *cdE* (r''), and *CDE* (R^z) were postulated to have arisen from the more common "chromosomes" *CDe* (R^1), *cDE* (R^2), and *cde* (r) by crossing over. For example, by rare crossing over, persons of genotype *CDe/cde* (R^1r) could produce the gametes *Cde* (r') and *cDe* (R^0); similarly, *cDE/cde* (R^2r) individuals could rarely produce gametes *cDe* (R^0) and *cdE* (r''), while persons of genotype R^1R^2 could produce gametes R^0 and R^z (cf. fig. 3). Since every time r', r'' or R^z could arise by crossing-over, R^0 would also be formed, the relationship $R^0 = r' + r'' + R^z$ would be expected to hold. Moreover, "chromosome" *CdE* (r^y) could arise by crossing over mainly from the rare genotypes *Cde/cdE* ($r'r''$) and *CDE/cde* (R^zr) so that the frequency of gene r^y could be expected to be extremely rare according to Fisher's hypothesis. In Mongololds, where "chromosome" *cde* (r) is virtually absent, the original chromosomes were postulated to have been *CDe* (R^1) and *cDE* (R^2), so that from individuals of genotype R^1R^2, by rare crossing-over, gametes R^z and R^0 could be expected to result in equal numbers.

In the original studies of Race et al. on 2,000 Englishmen (cf. table 40), the prediction seemed to be fulfilled, since the estimated frequency of "chromosome" R^0 was 2.57, while the sum of the frequencies of "chromosomes" r', r'', and R^z was $1.24 + 0.73 + 0.22 =$

* In 1950, Lawler,[174] in a study on the heredity of the Rh-Hr blood types commented, "No example of crossing over has been found in families tested by this Unit, nor do we know of any certain examples found elsewhere . . . . The work done by Dr. Alexander Wiener's laboratory and this Unit has established with certainty the manner of inheritance of the allelomorphs of Rh . . . ."

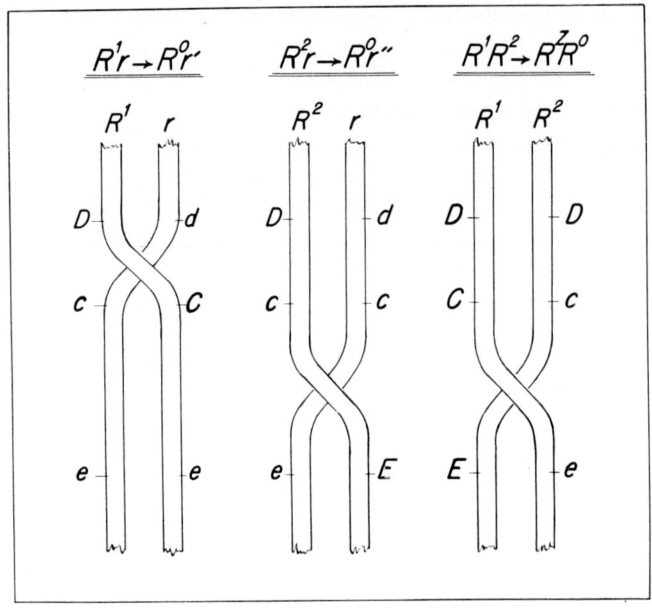

FIG. 3.—Fisher's Cross-over Hypothesis.

2.19, which is in reasonable agreement with the prediction. In addition, the report by Wiener et al. on Mexican Indians (cf. table 40) in which the frequency of the R^0 chromosome was 5.8 per cent and of the R^z "chromosome" 3.3 per cent was also considered to be in reasonable agreement with the expectations. Subsequently, the results of additional anthropological investigations became available and a number of these are summarized in table 40. As can be seen, the initial satisfactory agreement with the hypothesis has not been maintained in other populations, so that as Race and Sanger state in their book in 1954, "it is possible to pick results strikingly in favour of the crossing-over idea—and strikingly against it."

Fisher and Race have carried the idea further by estimating the distances between the hypothetical linked genes C-D-E, from the ratio of the frequency of each rare gene to that of the genotype from which it was postulated to have arisen. Based on the calculations for the British population, the genes were assumed to be arranged in the order *D-C-E*, and the distances between the genes were calculated to be such that $D\text{-}E = D\text{-}C + C\text{-}E$. As shown in table 40, however, when the same calculations were applied to other populations the relationship no longer held. One of the serious weaknesses of all these calculations is that no allowance was made for agglutinogens characterized by $\mathbf{Rh}_0$ variants or variants of other Rh-Hr factors.

Additional evidence considered to be in support of the theory of linked genes and of the arrangement *D-C-E* was the discovery of blood specimens having factor $\mathbf{Rh}_0$ (D) but lacking both members of the pairs *C-c* (**rh′-hr′**) and *E-e* (**rh″-hr″**). These bloods were designated as D- -/D- - to indicate that a deletion of the genes *C-c* and *E-e* from the chromosome had occurred. The similar example of Gunson and Donahue (cf. page 88) in which the members of the pair **rh″-hr″** (*E-e*) were missing has been designated DC^w-/DC^w-. Our own explanation for this situation, namely, the existence of additional alleles determining agglutinogens with super-$\mathbf{Rh}_0$ factors which crowd out the reactivity of other blood factors, is more in line with observations in other blood group systems such as the M-N-S types, where rare blood specimens have been encountered which lack the blood factors **S-s** as well as **U**. In this latter situation, postulating the existence of an additional allele has proved a satisfactory solution to the problem, without the need for hypothesizing the occurrence of a deletion in a chromosome. Furthermore, in Drosophila, for example, large deletions are lethal while small deletions are viable only in heterozygous form. In general, deletions produce marked phenotypic modifications; yet all individuals encountered to date who are homozygous for the supposed deletions *D- -* and DC^w- have been entirely normal.

The discovery of factor **hr** (f) made the six factor-six gene hypothesis completely untenable. This factor, as previously explained, occurs only in persons bearing the genes R^0 and/or r. The original explanation, namely, that this factor was present when c and e were present on the same chromosome but was not found when they were present together but on different chromosomes, the so-called position effect, was later changed to the postulation that a fourth locus existed, closely related to the C-D-E loci and occupied by a new gene pair *F-f* with anti-*F* yet to be found. [It may be pointed out that the postulated serum anti-*F* would be most difficult to recognize, since it should give reactions corresponding to the polyvalent serum anti-**rh′+rh″** (anti-$C+E$)]. As a result, the symbols for the various "chromosomes" had to be changed, so that for gene r the correspond-

TABLE 40

STATISTICAL TEST OF FISHER-RACE CROSS-OVER HYPOTHESIS

(After Wiener, A. S.: *Trans. N. Y. Acad. Sci.* 13, 199, 1951.)

Population	Investigator	Number of persons tested	Calculated frequencies (%) of genes ("chromosomes")							Calculated frequency (%) of genotypes			"Chromosome" segments		
			r	r'	r''	R^0	R^1	R^2	R^z	R^1R^2	R^1r	R^2r	*C-E* $\left(=\frac{R^z}{R^1R^2}\right)$	*D-C* $\left(=\frac{r'}{R^1r}\right)$	*D-E* $\left(=\frac{r''}{R^2r}\right)$
Caucasoids															
U. S. A.	Wiener *et al.*	2,390	36.65	1.23	0.52	3.73	42.70	15.06	.05	12.86	31.30	11.04	.004	.039	.045
U. S. A.	Unger *et al.*	7,317	38.3	1.4	0.8	2.8	42.1	15.1	—	12.71	32.25	11.57	0	.043	.069
England	Race *et al.*	2,000	28.86	0.98	1.19	2.57	42.05	14.11	0.24	11.87	32.68	10.97	.020	.030	.108
Canada	Chown *et al.*	3,100	39.55	1.24	0.73	1.91	43.48	12.87	0.22	11.19	34.39	10.18	.020	.036	.072
Spain	Race *et al.*	223	36.95	0	0.61	0.61	50.11	12.16	0.45	12.19	37.03	8.99	.037	0	.051
Czechoslovakia	Raska *et al.*	181	40.02	0.69	0.69	1.36	39.59	16.90	0.75	13.38	31.68	13.52	.056	.022	.051
Basques	Chalmers *et al.*	383	53.16	1.47	0.25	0.50	37.56	7.07	0	5.31	39.93	7.51	0	.037	.033
Negroids, N. Y. C.	Wiener *et al.*	223	28.4	2.7	0	42.1	11.7	14.4	0	3.37	6.65	8.18	0	.41	0
	Wiener *et al.*	200	27.39	0.89	0.89	43.32	17.30	9.02	0	3.10	9.54	4.89	0	.093	.182
Pygmies, Belgian Congo	Hubinont and Snoeck	94	10.5	0	0	63.2	6.2	19.5	0	2.4	1.3	4.1	0	0	0
Siamese	Phansanboom *et al.*	213	0	0	0	11.13	75.54	11.13	2.16	16.82	0	0	.128	—	—
Papuans	Simmons and Graydon	100	0	0	0	2.1	94.3	2.0	1.6	3.8	0	0	.42	—	—

Australian aborigines	Simmons and Graydon	234	0	12.87	0	8.54	56.42	20.09	2.08	22.67	0	0	.091	—	—
New Caledonians	Simmons *et al.*	325	0	0	0	5.48	83.32	10.77	0.43	17.95	0	0	.025	—	—
Mexican Indians	Wiener *et al.*	98	0	0	0	5.8	64.1	26.8	3.3	34.36	0	0	.096	—	—
Navaho Indians	Boyd and Boyd	305	0	17.31	2.02	8.37	31.05	35.33	5.73	21.9	0	0	.262	—	—

ing "chromosome" becomes *dcef*; $r' = dCeF$; $r'' = dcEF$; $R^0 = Dcef$, etc.

To date, no position within the four sets of linked genes has been assigned to the gene corresponding to the factor $\mathbf{hr}^{\mathrm{v}}$ or **V** discussed on page 81, and, similarly, no provision has been made in the scheme for factors $\mathbf{Rh}^{\mathrm{A}}$, $\mathbf{Rh}^{\mathrm{B}}$, $\mathbf{Rh}^{\mathrm{C}}$, ...

With the lack of evidence for crossing-over between the various postulated gene pairs *D-d*, *C-c*, *E-e*, and *F-f*, Race has invoked the concept of "complete" linkage. This latest alteration in the theory makes quite meaningless the concept of linear arrangement, since, operationally, the linked gene theory thus becomes identical with Wiener's theory of multiple alleles, making the substitutions $cdef = r$, $CdeF = r'$, etc. The populations in which the relationship $R^0 = R^z + r' + r''$ appears to hold so well strongly suggest that Fisher's suggestion that in Caucasoids the more common genes R^1, R^2 and r may have been the "original" genes may be correct. However, the rarer genes r', r'', R^z, and r^y probably arose, not by crossing-over, but by mutation. If mutation occurred in steps, as seems likely, this could account for the "intermediate" genes corresponding to the various Rh-Hr variants, which are not taken into account by Fisher's original hypothesis. Similarly, in Mongoloids the "original" genes appear to have been R^1 and R^2; in Negroids the original genes appear to have been principally R^0 with some R^1 and R^2, accounting for the relatively high incidence of $\mathbf{Rh}_0$ variants in such populations.

The C-D-E Notations. From the foregoing, it is clear that the C-D-E notations differ fundamentally from the Rh-Hr nomenclature.[177]

1) Implicit in the C-D-E notations is the concept that each Rh-Hr agglutinogen has but a single blood factor with a single corresponding antibody, making it unnecessary to make any distinction between blood factor and agglutinogen. For example, instead of the agglutinogen Rh_1, a set of four agglutinogens *C*, *D*, *e*, *F* is visualized. As has been pointed out, however, a fundamental principle in immunology is that antigens have multiple corresponding antibodies and thus are characterized by multiple serological factors. This applied not only to the agglutinogens of the human blood groups, but also to the blood groups of cattle, fowl, mice, and even to the antigens of bacteria and viruses, and soluble proteins. The Rh-Hr nomenclature with its distinctive type faces for agglutinogen and blood factors automatically allows for this fundamental serological reality.

2) The C-D-E-F notations are based on the concept of inheritance of the agglutinogens in uniform sets of three, or four, if the postulated F-f pair is included. Actually, the number of serological factors which characterize an agglutinogen is almost unlimited, because of the complex character of the reacting sites, and thus varies from agglutinogen to agglutinogen, as studies on the Rh-Hr agglutinogens have shown (cf. table 34). In this respect, the Rh-Hr nomenclature is more nearly realistic.

3) The Rh-Hr terminology takes into account the existence not only of sharply differentiated blood factors, but also of graded series of related blood factors, the latter being indicated by varying the type face, as in the case of the $\mathbf{Rh_0}$ variants. The C-D-E notations treats the two cases in the same manner. As a result, while the blood factors C^w, C^x have their corresponding specific antiserums anti-C^w, anti-C^x, certain variants of $\mathbf{Rh_0}$ are indicated by the analogous symbol D^u even though there is no corresponding anti-D^u serum.

4) A logical sequel of the concept of a one-to-one correspondence between antigen and antibody is the interpretation of certain serums as containing "linked" antibodies.[178] For example, in the case of the A-B-O system the anti-**C** antibody had to be designated anti-$A+B$ even though the more reasonable interpretation of the reaction of such antiserum is that it identifies an additional serological factor of the red cells, namely, blood factor **C**. Similarly, in the Rh-Hr system linked antibodies anti-$C+C^w$, had to be postulated to explain the reactions of anti-**rh′** serum with bloods Rh_1^w (DC^weF) as well as Rh_1 (DCeF).

5) The C-D-E notations pretend to indicate in the symbols all of the reactions of the red cell, and are therefore presumed to be self-explanatory. The Rh-Hr nomenclature makes no attempt to include all the serological attributes of the agglutinogen in the symbols, which instead are in the nature of simple mnemonics, as has already been explained. Of interest, are the contrasting effects on the symbols of the discovery of anti-*f* (anti-**hr**) and anti-*V* (anti-$\mathbf{hr^v}$) serums. While the Rh-Hr symbol of gene *r* remained essentially unchanged, the corresponding C-D-E symbol *cde* had to be expanded first to *cdef* and then to *cdefv*. Moreover, it has proved impossible to expand the C-D-E notations to keep abreast of the newer discoveries, such as $\mathbf{Rh^A}$, $\mathbf{Rh^B}$, $\mathbf{Rh^C}$ etc. In fact, in some instances where the discoveries have not followed the theoretical predictions, symbols for

certain factors have been included in the notations even though the factors have not yet been proved to exist, as in the case of *d* and *F*.[179]

6) A serious difficulty with the C-D-E-F notations is the lack of agreement in the use of these symbols by different workers. This is particularly true of the symbols for the various phenotypes. For example,[180] the phenotype Rh_1rh has been variously designated as CcDee; CDe/cde; (+++−); (++−+); C+ c+ D+ E− e+; DCe/−ce; DCeF/dcef; C(A+ B+),D(A− B+),E(A− B+),F(A+); etc. etc.

7) Instead of reporting phenotypes many workers who use the C-D-E symbols report "most likely genotypes". In fact, the phrase "most likely" is frequently omitted and "genotypes" are reported based on serological tests alone. This procedure introduces significant statistical errors, and in a number of instances has resulted in confusion. As has already been pointed out, the Rh-Hr nomenclature has distinctive symbols for phenotypes and genotypes, which automatically avoid this pitfall.

The Committee on Medicolegal Problems of the American Medical Association has prepared a report[181] on the medicolegal application of blood grouping tests in which one of the most important problems discussed was that of Rh-Hr nomenclature. In the report, the recommendation was made that "unless and until some other convention can be agreed upon, the original Rh-Hr notations be retained as the standard and sole nomenclature for the preparation of approved medicolegal reports on Rh types." In view of the facts which have been presented here, it would appear that this recommendation should now be extended to include the applications of the Rh-Hr blood types in clinical medicine and anthropology.

In conclusion, it is necessary to reiterate that the essential difference between the two concepts is on a serological, and not a genetic level. Blood grouping is a subspecialty of immunology, and while a knowledge of genetics is helpful, it is only of secondary importance for mastery of the field. The linkage theory completely evades the fundamental immunological concept of the difference between an agglutinogen and its serological attributes, the blood factors, and therein lie the pitfalls of the C-D-E notations. Thus, the issue at stake is not merely a choice between two nomenclatures, but the accuracy in reporting scientific observation.

CHAPTER VI
The Kell Blood Groups

In 1946, Coombs et al.[182] encountered an antibody in the serum of the mother of an erythroblastotic baby which gave reactions unrelated to any previously described blood group system. The patient's serum clumped the red blood cells of approximately 7 per cent of all persons tested. The blood factor detected by this antibody was named **Kell** after the patient. Wiener and Gordon[183] independently discovered in the serum of a patient who had had a hemolytic transfusion reaction an antibody which reacted with blood samples of 12.9 percent of individuals tested. The factor was named **Si** after the patient, and in family studies was found to be inherited as a simple Mendelian dominant, apparently independent of the A-B-O, M-N-S, P, and Rh-Hr blood types. Comparison by Mourant[184] of the two antiserums anti-**Si** and anti-**Kell** showed them to be identical. For the sake of uniformity, therefore, the symbol **Si** was discarded by Wiener, and the symbol **K** is now used to designate the **Kell** or **Si** factor.

Subsequently, other investigators found further examples of this antibody, and the distribution of the **K** factor has been studied in many populations. Among Caucasoids, the incidence has been found to vary from a low of about 7 per cent to as high as 13 per cent. Among Negroids, the incidence is considerably lower, while among Mongoloids the **Kell** factor has not been found.

The most extensive studies have been done among Caucasoids. Of these, one of the largest series is that of Lewis, Chown et al.,[185] who observed among 2,872 Canadians 198 or 6.89 per cent positive reactors. If we assume that the **Kell** factor is transmitted as a simple Mendelian dominant the gene frequencies for this series can be readily estimated as follows:

$$k = \sqrt{\text{Kell neg}} = \sqrt{0.931} = 0.965$$

$$\text{and } K = 1 - 0.965 = 0.035.$$

where K is the dominant gene and k its allele.

The calculated frequencies of the three possible genotypes for this population then are

$$\left.\begin{array}{ll}\text{Genotype } KK & = (0.035)^2 = 0.001225 = 0.12\% \\ \text{Genotype } Kk & = 2(0.035)(0.965) = 0.6755 \\ & = 6.75\%\end{array}\right\}\begin{array}{l}\text{Kell} \\ \text{positives} \\ 6.87\%\end{array}$$

and

$$\text{Genotype } kk = (0.965)^2 = 0.9312 = 93.12\%$$

Additional family studies were subsequently carried out, particularly by British investigators, and Race and Sanger[1e] have summarized studies on 336 families with 743 children. Among 507 children resulting from the mating **K**− × **K**−, none were observed who were **K** positive.

Factor *k* (Cellano)

In 1949, Levine et al.[186] found in the serum of a mother of a mildly affected erythroblastotic baby an antibody, which when tested against 2,500 different blood samples clumped all but five. This high frequency of positive reactors suggested that the factor it detected might be **k**, determined by the allele of **K**, especially since all negative reactors were **Kell** positive (cf. table 41). The expected incidence of negative reactions in New York and London was about 1 in 500, in close agreement with the frequencies obtained by Levine et al. with their antiserum. Therefore, the antibody in question, originally designated **Cellano** after the patient, is now called anti-**k**. Additional evidence in favor of the theory was obtained by family studies. A particularly striking example is the family described by Levine *et al.* (table 42). Similar families have been described by Shapiro[187] and by Wiener.[188]

The series of 2,872 Canadians described above was tested by Lewis *et al.* for factor **k** as well as factor **K**. The distribution of the three K-k types observed by them was: type KK, 0.24%; type Kk, 6.65%; and kk, 93.11%. From these values the gene frequencies are readily calculated by direct count, as follows:

$$\text{gene } K = \text{KK} + \frac{\text{Kk}}{2} = 0.24 + \frac{6.65}{2} = 3.57\%$$

$$\text{gene } k = \text{kk} + \frac{\text{Kk}}{2} = 92.99 + \frac{6.88}{2} = 96.43\%$$

With these gene frequencies, the expected distribution of the three

TABLE 41
THE K-k BLOOD TYPES

Tests Done Only for Factor **K**		Tests Done for Both Factors **K** and **k**			
Phenotypes	Reaction with Anti-**K**	Phenotypes	Reactions with		Genotypes
			Anti-**K**	Anti-**k**	
K (or K+)	+	KK	+	−	*KK*
		Kk	+	+	*Kk*
k (or K−)	−	kk	−	+	*kk*

TABLE 42
FAMILY SHOWING GENETIC RELATIONSHIP
BETWEEN FACTORS **K** AND **k**
(After Levine *et al.*[186])

Blood of	Reactions with Serum		Phenotype
	Anti-**K**	Anti-**k**	
Mother	+	+	Kk
Father	+	+	Kk
Children 1.	+	+	Kk
2.	+	−	KK
3.	−	+	kk
4.	+	+	Kk
5.	+	−	KK
6.	+	−	KK
7.	+	+	Kk
8.	−	+	kk

K-k is: type KK = 0.13%; type Kk = 6.88%; and type kk = 92.99%. Therefore, $\chi^2 = 3.28$, and $.01 < P < .05$. The large value of chi square is due mainly to the excess of individuals of the rare type KK, namely, 7 instead of the expected 3.5 among the 2,674 persons tested. While this may well be accidental, another possibility is that some of the individuals classified as type KK were actually type Kk, but had red cells reacting weakly with anti-**k** serum. The latter interpretation seems to be supported by some recent observations of Allen *et al.* now to be described.

Additional K-k Factors

Allen and Lewis[189] encountered an antibody for a previously unknown blood factor in the serum of an individual who, as far as was known, had never been exposed to the antigen either through pregnancy or transfusion. Subsequently, two more examples of serums of identical specificity were found, again in donors who had not been sensitized by either transfusion or pregnancy. The new factor was named **Penney** after the first donor, and was assigned the symbol **Kp**^{a}. Among 2,363 individuals in the Boston area, 51 (2.16 per cent) were found to have the antigen, an approximate "gene" frequency of 1 per cent.

The **Penney** factor was shown by Allen and Lewis to be related to the K-k system by the following observations. (1) Among 73 **Penney** positive individuals, only 2 were type K, while the expected incidence of **K**+ individuals was 9.8 per cent. (2) In certain families, mothers who were apparently of type kk had children apparently of type KK, while some mothers who were typed as KK had children apparently of type kk.* In all such instances, the mothers and the children proved to be **Penney** positive, so that it became necessary to postulate the presence in these families of a third gene, allelic to K and k, which determined the Penney antigen. That is, the mother and child in these exceptional instances actually belonged to genotypes Kk^{P} and kk^{P} respectively, or vice versa.† (3) In matings in which one parent had both factors **K** and **Kp**^{a} and the other parent lacked both these factors, no child had both factors or lacked both factors; instead, half had factor **K** while half had factor **Kp**^{a}. By using an exceptionally potent anti-**k** serum, or by the use of the absorption technique, it was subsequently found that the **Penney** positive individuals also had the factor **k**, or a variant thereof, which was being missed by the anti-**k** serum used up to that time.

In 1956, Fudenberg[190] encountered an unusual antiserum in a man who had received several transfusions. This serum agglutinated the red blood cells of all but 1 of 2,363 unrelated persons. This single negatively-reacting blood was **Kell** negative, and **Penney** positive, and gave weak reaction with anti-**k**. The serum also failed to agglutinate the red cells of one of the patient's two brothers. Like the cases of the negatively reacting blood found in the random series, this blood also gave weak reactions with anti-**k**. Exceptionally strong reactions were obtained, however, with both the patient's blood cells and those of his brother when tested with anti-**Kp**^{a} serum, as one

* In this study, only **K**+ individuals were tested with anti-**k** serum; all **K**− persons were assumed to be type kk.

† The new allelic gene is assigned the symbol k^{P} instead of K^{P}, for reasons which will become apparent later.

might expect if these individuals were of genotype $k^P k^P$, and for this reason the symbol **Kp**^{b} was assigned to the new factor (also called Rautenberg after the patient). Allen and Lewis suggest that with the four antiserums anti-**K**, anti-**k**, anti-**Kp**^{a} and anti-**Kp**^{b}, probably 10 genotypes and 9 phenotypes can be distinguished. (There is convincing evidence, however, for only 3 alleles, *K*, *k*, and k^P, which can yield only 6 genotypes).

No final genetic interpretation is possible at the present time because of the limited facts available. It appears that the agglutinogen K, determined by gene *K* has, in addition to factor **K**, factor **Kp**^{b}. The gene formerly designated *k* seems to include at least two varieties. The more common, making up more than 98 per cent, for which the designation gene *k* may still be retained, determines an agglutinogen characterized not only by the factor **k**, but also factor **Kp**^{b}. The remainder, which may be assigned the symbol gene k^P, determines a different rare agglutinogen having factor **Kp**^{a}, and in this case the factor **k** is poorly defined, judging from the weaker reactions with anti-**k** serum. According to a report by Chown,[191] there may be, in addition, a fourth rare allele in this system, tentatively designated K^o, which determines an agglutinogen that has none of the blood factors **K**, **k**, **Kp**^{a} or **Kp**^{b}.

Serology

The factors **K** and **k** are far less antigenic than the **Rh**$_0$ factor, judging from immunization experiments and clinical observations.[188, 192, 193] As a cause of hemolytic transfusion reactions and erythroblastosis fetalis factor **K** ranks with factor **hr′** in importance. Unfortunately, however, the tests for the K-k factors are much more delicate than the tests for the Rh-Hr factors. Anti-**K** and anti-**k** are not reactive by the proteolytic enzyme technique; they may react by the saline agglutination or conglutination techniques; but the most satisfactory reactions are obtained with such reagents by the anti-globulin method. The clumps in positive reactions are more fragile than the agglutinates of Rh-Hr tests, apparently due to the smaller number of reacting sites on the red cell envelope,[194] and this makes the reactions reproducible with greater difficulty than other serological tests. Reliable results, therefore, are possible only by painstaking work carried out by experienced investigators, and these tests have not been sufficiently perfected to be used routinely in medicolegal cases of disputed parentage.

CHAPTER VII

The Lutheran Blood Groups

The first antiserum for a blood factor of the Lutheran system was described in 1946 by Callender and Race.[195] The antibody was produced by a patient with lupus erythematosus, whose serum contained a number of unusual antibodies as the result of a series of blood transfusions. The factor was named **Lutheran** after the donor whose blood sensitized the patient, and was assigned the symbol $\mathbf{Lu^a}$ while the corresponding antibody was designated anti-$\mathbf{Lu^a}$. A number of additional antiserums of specificity anti-$\mathbf{Lu^a}$ were later encountered by other workers, but the reagent is still rare, so that the number of investigations carried out with it is limited.

According to Greenwalt and Sasaki[196] the incidence of the Lu(a+) type in pooled studies of 2,539 Englishmen was calculated by Mourant to be 6.93 per cent. If heredity by a pair of allelic genes Lu^a and Lu^b is postulated, the genotype frequencies calculated from these data are: Lu^aLu^a, 0.12%; Lu^aLu^b, 6.81%; and Lu^bLu^b, 93.07%.

Family Studies. Callender and Race tested the bloods of relatives of 17 Lu(a+) persons and the results indicated that the $\mathbf{Lu^a}$ factor was transmitted as a Mendelian dominant. Moreover, the $\mathbf{Lu^a}$ factor appeared to be inherited independently of the A-B-O, M-N, Rh-Hr, P-p, and K-k systems.

Lawler[197] studied the heredity of the Lutheran groups in 47 families with 97 children while Race, Sanger and Thompson[198] tested 257 families with 594 children. The results conform with the genetic theory; in particular, in 254 families where both parents lacked the blood factor all of 601 children were likewise Lu(a−).

Mainwaring and Pickles[199] reported that their anti-$\mathbf{Lu^a}$ serum distinguished strongly and weakly reacting Lu(a+) cells, analogous to the subgroups of A, and postulated three corresponding allelic genes.

Mohr et al.[200] tested many large Danish families for the Lewis and Lutheran blood groups, and interpreted their results as evidence for linkage between the genes *Le* and *Lu*. However, considering the difficulties in carrying out the Lewis tests, as described on page 32, it would seem premature to accept these conclusions without confirmation.

Distribution of Lu^a. According to Mourant,[201] among European whites different investigators have reported the frequency of the Lu(a+) type to range between 3 and 10 per cent. In a limited number of studies carried out so far on natives of Asia and Australia the Lu(a+) types have been observed in widely varying frequencies. In Africa, frequencies varying from zero for Bushman to 17 per cent for Nigerians have been reported.

Factor Lu^b. Cutbush and Chanarin[202] found an antibody in the serum of a woman who had three normal pregnancies but no transfusions, which gave the reactions expected for anti-$\mathbf{Lu^b}$. Recently, Greenwalt and Sasaki[196] described another example of such a serum in a woman who was sensitized by a blood transfusion. To date, no extended anthropological or genetic studies have been carried out with serum of this specificity.

The Lutheran blood groups have not proved to be of any considerable clinical importance up to now, and remain primarily of academic interest.

CHAPTER VIII
The Duffy Blood Groups

In 1950, Cutbush, Mollison, and Parkin,[203] and van Loghem and Hart[204] independently reported the discovery of a blood factor, unrelated to the A-B-O, M-N-S, P-p, Rh-Hr and Kell blood types. The antibody described by Cutbush et al. was found in the serum of a hemophiliac who had sustained a mild transfusion reaction after a series of transfusions received over a period of twenty years. The blood factor was named **Duffy** after the patient. Other examples of sensitization to the **Duffy** factor were encountered by other investigators, and one of us (A. S. W.) encountered such a serum with a titer of more than 500 units in a woman who had a serious hemolytic transfusion reaction caused by sensitization to the **Duffy** factor

Terminology. In order to avoid confusion with the C-D-E symbols for the Rh-Hr types, instead of the letter *D*, the symbol *Fy* was selected by Cutbush et al. to represent the **Duffy** factor. Wiener[205] has suggested shortening this symbol to **F**, and this latter terminology will be used in the present discussion.

Serology. All the anti-**F** serums encountered up to the present time have been usable only by the anti-globulin technique. Obviously, therefore, when the red cells are already coated, as in cases of auto-hemolytic anemia or in erythroblastosis fetalis, the Duffy tests can not be carried out reliably. The clumps which characterize a positive reaction may be large, but they are quite friable, even more so than the clumps of the Kell tests. This Wiener[206] ascribes to the smaller number of antigenic sites on the red cell envelope, which he estimates to be about one-tenth as numerous as the antigenic sites of the Rh-Hr agglutinogens. Because of the delicate nature of the reactions, the results are duplicated with difficulty, and it is only with painstaking work that reliable results are possible.

Distribution. Among Caucasoids, the frequency of the **F**-positive individuals ranges between 60 and 70 per cent; among Negroids, the frequency is considerably lower, namely, between 10 and 25 per cent; while in Mongoloids the incidence is highest, in some populations approaching 100 per cent.

One of the largest series of persons examined to date is that reported by Ikin *et al.*,[207] who tested 1,116 Englishmen and found 65.52 per cent type F and 34.48 type f. From these values the gene frequencies are readily calculated as follows:

$$f = \sqrt{\text{type f}} = \sqrt{0.3448} = 58.72\%$$

$$F = 100 - f = 41.28\%$$

From these values, the frequencies of the three theoretically possible genotypes can be readily calculated.

$$\left.\begin{array}{lll} \text{Genotype } FF & = (0.4128)^2 & = 17.04\% \\ \text{Genotype } Ff & = 2(0.4128)(0.5872) & = 48.48\% \end{array}\right\} \text{type F} = 65.52\%$$

$$\text{Genotype } ff = (0.5872)^2 = 34.48\%$$

Heredity. The few family studies carried out to date support the hypothesis that the **Duffy** factor is transmitted as a simple Mendelian dominant. In the combined studies by Cutbush and Mollison,[203] and Race and Sanger[208] there are 34 families in which both parents were type f, and among the 78 children of these families none were type F.

Factor f. Theoretically, if gene f gives rise to an agglutinogen f with a corresponding blood factor **f**, the corresponding serum anti-**f** should react with blood specimens of genotypes ff and Ff, but not with those of genotype FF. In 1951, Ikin et al.[209] detected an antibody, during a routine post-natal examination, which in tests on blood specimens from 59 persons gave the reactions expected for anti-**f**. Unfortunately, further studies were prevented because the serum lost its activity. In more recent studies, R. Sanger, has found that among Negroes blood lacking both factor **F** and **f** are not uncommon, so that the situation is more complex than it seemed at first.

CHAPTER IX

The Kidd Blood Groups

This blood group system was first described in 1951 by Allen, Diamond and Niedziela,[210] who found the new antibody in the serum of the mother of an erythroblastotic baby. The blood factor detected by the antibody was named **Kidd** after the patient. It was found to be independent of the A-B-O, M-N-S, P-p, Rh-Hr, Kell, Duffy, and Lutheran systems.

Nomenclature. In order to avoid confusion with the symbol for the **Kell** factor, the **Kidd** factor has been assigned the symbol *Jk* by Allen et al. Wiener[205] has suggested shortening the symbol to **J**, which is the symbol used in this discussion.

Distribution. According to a summary compiled by Mourant,[201] among Caucasoids the incidence of factor **J** is approximately 75 per cent. Among Negroids, the incidence has been found to be above 90 per cent, while among Mongoloids the frequency has ranged from 50 to 100 per cent. The largest series is that of Rosenfield et al.,[211] who found 76.72 per cent type J and 23.28 per cent type j among New York whites. From these figures the theoretical gene frequencies are readily determined as follows.

$$j = \sqrt{\text{type j}} = \sqrt{0.2328} = 48.25\%$$

$$J = 100 - 48.25 = 51.75\%$$

And from these estimated gene frequencies, the distribution of the three theoretically possible genotypes can be calculated.

$$\left.\begin{array}{lll} \text{Genotype } JJ & = (0.5175)^2 & = 26.78\% \\ \text{Genotype } Jj & = 2(0.5175)(0.4825) & = 49.94\% \end{array}\right\} \text{Type J} = 76.72\%$$

$$\text{Genotype } jj = (0.4845) = 23.28\%$$

Factor j. If the gene *j* gives rise to a corresponding agglutinogen j, the blood factor **j** should be present in the blood cells of all individuals of genotype *Jj* and *jj*. According to Race and Sanger, at least three antiserums have been encountered giving reactions expected for anti-**j**.

Serology. The anti-**J** serums described in the literature, as well as those encountered by one of us (A. S. W.), have been usable only by the anti-globulin technique. Therefore, these tests have the same limitations as the Duffy tests. Under ideal conditions, the reactions are clear-cut and readily visible to the naked eye, but the agglutinates are friable like those obtained with Duffy antiserums. Another difficulty is the tendency of the antiserums to lose their activity upon storage. Therefore, reliable results are obtained only by carefully repeating the tests several times, using the blind technique, and including controls of known types J and j.

CHAPTER X
Other Blood Group Systems

In addition to the blood factors which have been discussed, many others have been reported. In this section we propose to describe those additional blood factors which have not been shown to be related to the A-B-O, Lewis, M-N-S, P, Rh-Hr, Kell, Duffy, or Kidd systems. It is quite possible that some or even all of the factors described here may eventually prove to be related to one of the already known blood group systems, and it is also possible that there may be duplications, since complete comparative studies have not been practicable in view of the large number of blood factors involved. The chief difficulty in studying the factors in question lies in their distribution, since some are of almost universal occurrence while others are extremely rare. However, an agglutinogen which is extremely rare in one ethnic group may prove to be quite common in another population. For example, the **Kell**-positive blood type which is not uncommon among Caucasoids has yet to be demonstrated in Mongoloids. An example of the reverse is the Diego blood type which will now be described.

Diego System. The **Di** antibody was first found by Levine et al.[212] in the serum of the mother of an erythroblastotic baby, whose family name Diego was adopted as the designation for the blood factor. Examination of the family of the propositus showed the factor to be inherited as a simple Mendelian dominant, but tests on 2,600 consecutive Caucasoid blood specimens failed to show the presence of the **Diego** factor.[213] These studies were continued by Layrisse and Arends[214, 215] and by Lewis et al.[216] who tested blood from American Indians, among whom the **Diego** factor was found to be common. In different American Indian tribes the frequency of the **Diego** factor has been found to range from 2 per cent to 45 per cent. Layrisse and Arends[217] also found the factor to be common among Chinese and Japanese, and in pure-blood Negroes, Eskimos and Polynesians. Layrisse et al. have recorded the factor in three South American Indian families and confirmed its inheritance as a simple Mendelian dominant. Lewis et al.[216] have reported the inheritance of the **Diego** factor in 50 Japanese families in Canada. In 40 of the families both parents were **Diego** negative, and in none of the 111 children did the factor appear.

High Frequency Blood Factors. As examples of blood factors with a high frequency in the general population may be mentioned the factor **U** of the M-N-S system, factor **H** of the A-B-O system, factor **hr″** of the Rh-Hr system, **Tj** of the P system, and factors **k** and **Kp**^{b} of the Kell system. In addition to these, at least three other high frequency blood factors are known, namely, **Vel**, **I**, and **Yt**^{a} (cf. table 1). Of these three factors, factor **I** has been studied most intensively and is of the greatest interest.

Anti-**I** was found[218] in the serum of a group A_1B woman with acquired hemolytic anemia, whose serum also contained cold autoagglutinins in high titer. When this patient had hemolytic reactions following transfusions, her serum was also shown to contain an isoantibody of high titer, anti-**I**, reactive at room temperature with almost all other human bloods but not with the patient's own cells. No other member of the patient's family was found to be **I** negative (type i). Among a series of 22,964 random donors only 5 proved to be **I** negative, and of these 4 were Negroids. Among the positively-reacting blood specimens marked differences in titer were obtained, ranging from 25 to 5,000 units, thus indicating the existence of variants of the I agglutinogen, namely, I_1, I_2, I_3, . . . i. Recently, Shapiro[219] has encountered a second example of anti-**I** in South Africa, although of lower titer. In this case both the propositus and her child proved to be **I** negative.

TABLE 43

SOME LOW FREQUENCY BLOOD FACTORS

Name of Blood Factor	Observers	Random Series		Family of Propositus	
		+	−	+	−
Levay	Callender and Race (1946)	0	350	3	4
—	Wiener (1942)	0	ca. 500	1	3
Jobbins	Gilbey (1947)	0	120	2	2
Becker	Elbel and Prokop (1951)	0	272	3	4
Berrens	Davidsohn et al. (1953)	0	448	4	3
Cavaliere*	Wiener and Brancato (1953)	0	48	4	3

* F. H. Allen has recently shown that this blood factor is probably identical with the Wright factor (**Wr**^{a}), described at about the same time by Holman (*Lancet* **2:** 119, 1953).

Low Frequency Blood Factors. As examples of the low frequency factors may be mentioned **Mi** and **Gr** of the M-N-S system, $\mathbf{rh}^{w}$ and $\mathbf{rh}^{x}$ of the Rh-Hr system, "super" $\mathbf{Rh}_0$ of the Rh-Hr system, and $\mathbf{Kp}^{a}$ of the Kell system. In addition to these, a large number of blood factors have been reported, generally detected with the serum of mothers of erythroblastotic babies. Due to the low frequency of these factors the genetic investigations have been limited to the families of the propositi. The findings in each are compatible with heredity as a simple Mendelian dominant. The more fully described of these blood factors are listed in table 43.

As an example of a low frequency factor may be mentioned the case report of Wiener and Brancato.[222] A woman who had babies with erythroblastosis caused by $\mathbf{Rh}_0$ sensitization, then had an erythroblastotic baby who proved to be $\mathbf{Rh}_0$ negative. Tests of the maternal serum demonstrated that in addition to anti-$\mathbf{Rh}_0$ there was another antibody present, which was designated anti-**Ca** after the patient. Among 7 $\mathbf{Rh}_0$-negative relatives the blood of 4 proved to have the factor **Ca**, while 48 consecutive $\mathbf{Rh}_0$-negative donors were all **Ca** negative. Another striking example is that of Davidsohn *et al.*[221] who found a potent antibody for a blood factor present in 4 out of 7 members of their patient's family, but absent from 448 consecutive blood donors (cf. table 43).

CHAPTER XI

Medicolegal Applications

The clear-cut hereditary transmission of the human blood group factors has led to their application for the solution of medicolegal problems of disputed parentage. The types of cases in which the tests have been applied are as follows:

1) A man accused of the paternity of a child born out of wedlock denies the charge. This is the most common type of case with which courts are confronted where blood grouping tests are used.
2) A man is accused of assault or rape, and tries to prove innocence by showing that he is not the father of the child born to the complainant.
3) A child is born in lawful wedlock, but the husband denies paternity.
4) It is suspected that two newborn infants have been interchanged in a nursery in a hospital.
5) A woman may simulate pregnancy and childbirth, and may claim that a certain child is hers to compel a man to marry her or to obtain dower in a dead husband's estate.
6) An individual who wished to immigrate to the U.S.A. asserts that his parents are American citizens living in the states, but it is suspected that the claim is fraudulent.[253]

A number of reports[222, 223, 224] regarding the medicolegal applications of blood grouping tests have been published by the Committee on Medicolegal Problems of the American Medical Association, and they should be consulted by those interested in this special problem. The legal aspects of the subject are discussed in detail in the excellent book by S. B. Schatkin.[225]

Blood tests can be used only to exclude and not to prove paternity. In certain jurisdictions laws have been passed empowering the courts to order blood grouping tests in cases where parentage is an issue. Naturally, the chances of disproving paternity will depend upon how complete an examination is carried out. However, the examination should be restricted to only those blood tests which have been proved to be reliable, and where an adequate number of heredity studies have been carried out. Therefore, the Committee on Medicolegal

Problems of the American Medical Association has recommended that the routine tests for court cases be limited to the **A** and **B** factors of the A-B-O system, the **M** and **N** factors of the M-N system, and the factors **Rh₀**, **rh′**, **rh″**, and **hr′** of the Rh-Hr system. In private cases, or under special circumstances, qualified medicolegal experts may find it desirable to carry out more complete examinations, which will vary depending on the experience of the investigator and on the reagents available to him.

For convenience of reference, simplified tables which show the application of the A-B-O, M-N, and Rh-Hr tests for excluding paternity have been prepared (cf. tables 44, 45, 46).

Chances of Excluding Paternity

If a man falsely accused of paternity is found to have a blood group genetically incompatible with that of the child this establishes his

TABLE 44

EXCLUSION OF PATERNITY BY THE A-B-O GROUPS*

Phenotype of the Putative Mother	Phenotype of Putative Father			
	O	A	B	AB
O	A, B, **AB**	B, **AB**	A, **AB**	O, **AB**
A	B, AB	B, AB	None	O
B	A, AB	None	A, AB	O
AB	**O**, AB	**O**	**O**	**O**

* Find the phenotypes of the putative father and mother at the top and side columns of the tables and locate the box at which these intersect. In the box are given the groups not possible in children of the mating; groups given in bold face type represent children for whom maternity is excluded.

TABLE 45

EXCLUSION OF PATERNITY BY THE M-N TYPES*

Phenotype of Putative Mother	Phenotype of Putatitive Father		
	M	N	MN
M	MN, **N**	M, **N**	**N**
N	**M**, N	**M**, MN	**M**
MN	N	M	None

* Find the phenotypes of the putative father and mother at the top and side columns of the table, and locate the box at which these intersect. In the box are given the types not possible in children of the mating; types given in bold face type represent children for whom maternity is excluded.

TABLE 46

EXCLUSION OF PATERNITY BY THE Rh-Hr TYPES

(Modified after W. C. Boyd)

Phenotype of Putative Mother	Phenotype of Putative Father											
	1 rh	2 rh'rh	3 rh'rh'	4 rh"	5 rh'rh"	6 $rh_y rh'$	7 Rh_0	8 $Rh_1 rh$	9 $Rh_1 Rh_1$	10 Rh_2	11 $Rh_1 Rh_2$	12 $Rh_Z Rh_1$
1. rh	2, **3**, 4, 5, **6**, 7, 8, **9**, 10, 11, **12**	**3**, 4, 5, **6**, 7, 8, **9**, 10, 11, **12**	1, **3**, 4, 5, **6**, 7, 8, **9**, 10, 11, **1**	2, **3**, 5, **6**, 7, 8, **9**, 10, 11, **12**	**3**, **6**, 7, 8, **9**, 10, 11, **12**	1, **3**, 4, **6**, 7, 8, **9**, 10, 11, **12**	2, **3**, 4, 5, **6**, 8, **9**, 10, 11, **12**	**3**, 4, 5, **6**, **9**, 10, 11, **12**	1, **3**, 4, 5, **6**, 7, **9**, 10, 11, **12**	2, **3**, 5, **6**, 8, **9**, 11, **12**	**3, 6, 9, 12**	1, **3**, 4, **6**, 7, **9**, 10, **12**
2. rh'rh	3, 4, 5, 6, 7, 8, 9, 10, 11, 12	4, 5, 6, 7, 8, 9, 10, 11, 12	1, 4, 5, 6, 7, 8, 9, 10, 11, 12	3, 6, 7, 8, 9, 10, 11, 12	7, 8, 9, 10, 11, 12	1, 4, 7, 8, 9, 10, 11, 12	3, 4, 5, 6, 9, 10, 11, 12	4, 5, 6, 10, 11, 12	1, 4, 5, 6, 7, 10, 11, 12	3, 6, 9, 12	none	1, 4, 7, 10
3. rh'rh'	**1**, 3, **4**, 5, 6, **7**, 8, 9, **10**, 11, 12	**1**, **4**, 5, 6, **7**, 8, 9, **10**, 11, 12	**1**, 2, **4**, 5, 6, **7**, 8, 9, **10**, 11, 12	**1**, 3, **4**, 6, **7**, 8, 9, **10**, 11, 12	**1**, **4**, **7**, 8, 9, **10**, 11, 12	**1**, 2, **4**, 5, **7**, 8, 9, **10**, 11, 12	**1**, 3, **4**, 5, 6, **7**, 9, **10**, 11, 12	**1**, **4**, 5, 6, **7**, **10**, 11, 12	**1**, 2, **4**, 5, 6, **7**, 8, **10**, 11, 12	**1**, 3, **4**, 6, **7**, 9, **10**, 12	**1, 4, 7, 10**	**1**, 2, **4**, 5, **7**, 8, **10**, 11
4. rh"	2, **3**, 5, **6**, 7, 8, **9**, 10, 11, **12**	**3**, **6**, 7, 8, **9**, 10, 11, **12**	1, **3**, 4, **6**, 7, 8, **9**, 10, 11, **12**	2, **3**, 5, **6**, 7, 8, **9**, 10, 11, **12**	**3**, **6**, 7, 8, **9**, 10, 11, **12**	1, **3**, 4, **6**, 7, 8, **9**, 10, 11, **12**	2, **3**, 5, **6**, 8, **9**, 11, **12**	**3, 6, 9, 12**	1, **3**, 4, **6**, 7, **9**, 10, **12**	2, **3**, 5, **6**, 8, **9**, 11, **12**	**3, 6, 9, 12**	1, **3**, 4, **6**, 7, **9**, 10, **12**
5. rh'rh"	3, 6, 7, 8, 9, 10, 11, 12	7, 8, 9, 10, 11, 12	1, 4, 7, 8, 9, 10, 11, 12	3, 6, 7, 8, 9, 10, 11, 12	7, 8, 9, 10. 11, 12	1, 4, 7, 8, 9, 10, 11, 12	3, 6, 9, 12	none	1, 4, 7, 10	3, 6, 9, 12	none	1, 4, 7, 10
6. $rh_y rh'$	**1**, 3, **4**, 6, **7**, 8, 9, **10**, 11, 12	**1**, **4**, **7**, 8, 9, **10**, 11, 12	**1**, 2, **4**, 5, **7**, 8, 9, **10**, 11, 12	**1**, 3, **4**, 6, **7**, 8, 9, **10**, 11, 12	**1**, **4**, **7**, 8, 9, **10**, 11, 12	**1**, 2, **4**, 5, **7**, 8, 9, **10**, 11, 12	**1**, 3, **4**, 6, **7**, 9, **10**, 12	**1**, **4**, **7**, **1**	**1**, 2, **4**, 5, **7**, 8, **10**, 11	**1**, 3, **4**, 6, **7**, 9, **10**, 12	**1, 4, 7, 10**	**1**, 2, **4**, 5, **7**, 8, **10**, 11
7. Rh_0	2, **3**, 4, 5, **6**, 8, **9**, 10, 11, **12**	**3**, 4, 5, **6**, **9**, 10, 11, **12**	1, **3**, 4, 5, **6**, 7, **9**, 10, 11, **12**	2, **3**, 5, **6**, 8, **9**, 11, **12**	**3, 6, 9, 12**	1, **3**, 4, **6**, 7, **9**, 10, **12**	2, **3**, 4, 5, **6**, 8, **9**, 10, 11, **12**	**3**, 4, 5, **6**, **9**, 10, 11, **12**	1, **3**, 4, 5, **6**, 7, **9**, 10, 11, **12**	2, **3**, 5, **6**, 8, **9**, 11, **12**	**3, 6, 9, 12**	1, **3**, 4, **6**, 7, **9**, 10, **12**
8. $Rh_1 rh$	3, 4, 5, 6, 9, 10, 11, 12	4, 5, 6, 10, 11, 12	1, 4, 5, 6, 7, 10, 11, 12	3, 6, 9, 12	none	1, 4, 7, 10	3, 4, 5, 6, 9, 10, 11, 12	4, 5, 6, 10, 11, 12	1, 4, 5, 6, 7, 10, 11, 12	3, 6, 9, 12	none	1, 4, 7, 10
9. $Rh_1 Rh_1$	**1**, 3, **4**, 5, 6, **7**, 9, **10**, 11, 12	**1**, **4**, 5, 6, **7**, **10**, 11, 12	**1**, 2, **4**, 5, 6, **7**, 8, **10**, 11, 12	**1**, 3, **4**, 6, **7**, 9, **10**, 12	**1, 4, 7, 10**	**1**, 2, **4**, 5, **7**, 8, **10**, 11	**1**, 3, **4**, 5, 6, **7**, 9, **10**, 11, 12	**1**, **4**, 5, 6, **7**, **10**, 11, 12	**1**, 2, **4**, 5, 6, **7**, 8, **10**, 11, 12	**1**, 3, **4**, 6, **7**, 9, **10**, 12	**1, 4, 7, 10**	**1**, 2, **4**, 5, **7**, 8, **10**, 11
10. Rh_2	2, **3**, 5, **6**, 8, **9**, 11, **12**	**3, 6, 9, 12**	1, **3**, 4, **6**, 7, **9**, 10, **12**	2, **3**, 5, **6**, 8, **9**, 11, **12**	**3, 6, 9, 12**	1, **3**, 4, **6**, 7, **9**, 10, **12**	2, **3**, 5, **6**, 8, **9**, 11, **12**	**3, 6, 9, 12**	1, **3**, 4, **6**, 7, **9**, 10, **12**	2, **3**, 5, **6**, 8, **9**, 11, **12**	**3, 6, 9, 12**	1, **3**, 4, **6**, 7, **9**, 10, **12**
11. $Rh_1 Rh_2$	3, 6, 9, 12	none	1, 4, 7, 10	3, 6, 9, 12	none	1, 4, 7, 10	3, 6, 9, 12	none	1, 4, 7, 10	3, 6, 9, 12	none	1, 4, 7, 10
12. $Rh_Z Rh_1$	**1**, 3, **4**, 6, **7**, 9, **10**, 12	**1, 4, 7, 10**	**1**, 2, **4**, 5, **7**, 8, **10**, 11	**1**, 3, **4**, 6, **7**, 9, **10**, 12	**1, 4, 7, 10**	**1**, 2, **4**, 5, **7**, 8, **10**, 11	**1**, 3, **4**, 6, **7**, 9, **10**, 12	**1, 4, 7, 10**	**1**, 2, **4**, 5, **7**, 8, **10**, 11	**1**, 3, **4**, 6, **7**, 9, **10**, 12	**1, 4, 7, 10**	**1**, 2, **4**, 5, **7**, 8, **10**, 11

1. Determine the phenotypes of the blood of the mother, putative father, and the child or children using potent and specific anti-**Rh_0**, anti-**rh'**, anti-**rh"**, and anti-**hr'** sera including appropriate controls.
2. Assign the corresponding phenotype symbol and number to each individual tested.
3. Find the assigned number of the mother and father on the side and top columns of the Table and locate the box at which both intersect.
4. The child or children are not possible from this mating if their assigned numbers appear in this box.
5. Phenotype numbers given in bold face represent children for whom maternity is excluded.

innocence. If the blood groups satisfy the laws of heredity this does not necessarily prove that he is the father, because of the limited number of types known and the possibility of coincidence. The chances of exonerating a falsely accused man will vary, depending on the distribution of the blood groups in the population, as well as on the blood group of the accused man. For example, in certain tribes of American Indians, where all of the members belong to group O, the A-B-O groups would be worthless for solving such problems.

Wiener[226, 227] has derived general formulae for the chances of excluding paternity by the A-B-O blood groups. These are given below, and those interested in their derivation should consult the original papers.

Blood group of falsely accused man	Chance of excluding paternity	
Group O	$r^2(p + q) + 2pq(1 + r)$	(42)
Group A	$q(p + r)^2$	(43)
Group B	$p(q + r)^2$	(44)
Group AB	r^2	(45)
Group undetermined	$p(q + r)^4 + q(p + r)^4 + pqr^2(p + q) + 2pqr^2$	(46)

In the above formulae p, q, and r represent the frequencies of genes I^A, I^B, and I^O, respectively. The maximum chance of excluding paternity is found to be approximately 20 per cent, and occurs when the gene frequencies are $p = 0.2213$, $q = 0.2213$ and $r = 0.5574$.

Similarly, formulae have been derived for the chances of excluding paternity, using the M-N types, as follows.

Blood type of falsely accused man	Chance of excluding paternity	
Type M	$n(1 - mn)$	(47)
Type N	$m(1 - mn)$	(48)
Type MN	0	(49)
Type undetermined	$mn(1 - mn)$	(50)

In the above formulae, m and n represent the frequencies of genes L^M and L^N, respectively. Here the maximum chance of excluding paternity is 18.75 per cent, and occurs when the gene frequencies are $m = 0.5$ and $n = 0.5$.

When the formulae are applied to populations such as that encountered in New York City table 45 could be drawn up. In making the calculations, account must be taken of the independent inheritance of the A-B-O and M-N types. If P_1 is the chance of excluding paternity by one blood group system and P_2 is the chance of excluding paternity by a second independent system, then the chances of exclusion when both are tested for is given by the formula

$$P = 1 - (1 - P_1)(1 - P_2). \quad (51)$$

This formula can be extended to include any number of blood group systems. It is of interest that the combined chances of excluding paternity by the combined use of the A-B-O and M-N tests in New York City is about 30 per cent, which is close to the theoretical maximum of 35 per cent.

General formulae for the chances of excluding paternity by the means of the Rh-Hr types have not been derived. These would be expected to be exceedingly complex. Wiener[228, 229] has estimated that the chances of excluding a falsely accused man of undetermined Rh-Hr type is close to 25 per cent, when tests are done for the factors **Rh**$_0$, **rh′**, **rh″**, and **hr′**. Thus, when tests are done for the factors mentioned of all of the three systems, the chances of excluding a falsely accused man is about 50 per cent. In practice the rate of exclusion is considerably less than 50 per cent, since most of the defendents in paternity proceedings will be the actual fathers. Mr. S. B. Schatkin and Mr. Sol Cooperman of the Corporation Council's Office of New York City have observed over a period of many years that approximately 17 or 18 men out of every 100 men who deny paternity are actually excluded by the tests, indicating that approximately 35 per cent of such accusations are false. In problems of disputed paternity involving married couples, as in divorce actions, our experience has been that the exclusion rate is considerably higher.

In selected cases it may be felt desirable to carry out additional tests, for example for the subgroups of A, secretor type, **P**, **S**, **hr″**, **Kell**, **Duffy** and other factors for which the expert has reliable antiserums available.

The formula for the chance of excluding paternity, using only a single blood factor **D**, transmitted as a Mendelian dominant, is given by the formulae below.

Blood type of falsely accused man	Chance of exclusion	
D+	0	(51)
D−	dr^2	(52)
Type undetermined	dr^4	(53)

Here d represents the frequency of the dominant gene, and r the frequency of the recessive gene. The maximum chance of excluding paternity is 0.08192 or about 8.2 per cent, and occurs when the gene frequencies are $d = 0.2$ and $r = 0.8$, respectively. In practice, it turns out that the distribution of most of the blood factors inherited in this way is such that the chance of excluding paternity is less than the maximal theoretical value. Moreover, because of the overlapping among tests, and the occurrence of double, triple, or quadruple exclusions, it becomes progressively more difficult to increase the chances of excluding paternity by the introduction of new blood tests. For example, a blood factor with an 8 per cent chance of exclusion will add only 4 per cent to the chances of exclusion, since one-half of the innocent men will already have been excluded by the A-B-O, M-N, and/or Rh-Hr tests. It has been estimated that all of the new blood tests combined add only about 12 to 15 per cent to the chances of excluding parentage, and it is questionable whether it is practicable or worth-while to carry out these difficult tests in such cases, since the results may not prove acceptable to the courts.

Boyd[230] has calculated in detail the chances of excluding paternity by the Rh-Hr tests in a Caucasoid population. His results, including the distribution of the Rh-Hr types, for the population to which his figures apply are given in table 48. For example, a falsely accused man of type rh has an almost 45 per cent chance of being excluded by tests for the Rh-Hr types. By combining the figures for tables 47 and 48, it is possible to calculate the chances of excluding paternity for a falsely accused man of any A-B-O group, M-N type, and Rh-Hr type. As an example, the chances for a group AB, type M, type Rh_1Rh_1 man would be $1 - (1 - 0.607)(1 - 0.4448) = 0.782$ or 78.2 per cent.

TABLE 47

CHANCES OF PROVING NON-PATERNITY WITH THE AGGLUTINOGENS A, B, M AND N*

(From Wiener's *Blood Groups and Transfusion.*)

Group......	O			A			B			AB			Unknown
Type.......	M	N	MN	M	N	MN	M	N	MN	M	N	MN	
Chances (Per cent)	50.0	54.6	23.5	39.6	45.1	7.7	44.1	49.3	14.6	60.7	63.4	39.9	31.9

* Based on the frequencies of the blood types in New York City.

TABLE 48

CHANCES OF EXCLUDING PATERNITY BY THE Rh-Hr TYPES

(Modified from Boyd, *Amer. J. Human Genet.*, **7:** 229, 1955)

Phenotype of Accused Man*	Frequency in Caucasoids (per cent)	Chance of Exclusion (per cent)
rh	14.36	44.95
rh′rh	0.62	18.45
rh′rh′	0.01	51.62
rh″	1.31	36.28
rh′rh″	0.03	9.73
Rh_0	2.41	44.14
Rh_1rh	35.77	10.66
Rh_1Rh_1	19.73	44.48
Rh_2	12.66	33.55
Rh_1Rh_2	12.99	0
Rh_ZRh_1	0.11	41.75
Unknown		25.0

* Not included in the table is the rare phenotype rh_yrh'.

Disputed Maternity

The blood tests may also be used to refute false claims of maternity. For example, maternity would be excluded if any of the following blood type combinations is encountered.

1) Mother group AB—child group O
   Mother group O—child group AB
2) Mother type M—child type N
   Mother type N—child type M
3) Mother type rh′rh′, rh_yrh', Rh_1Rh_1, or Rh_ZRh_1—child type rh, rh″, Rh_0, or Rh_2.
   Mother type rh, rh″, Rh_0, or Rh_2—child type rh′rh′, rh_yrh', Rh_1Rh_1, or Rh_ZRh_1.

The chances of excluding maternity are readily calculated as follows: For the A-B-O blood groups = $2(\overline{O})(\overline{AB}) = 2(0.40)(0.05)$ or approximately 4 per cent. For the M-N types = $2(\overline{M})(\overline{N}) = 2(0.30)(0.20)$ or approximately 12 per cent. For the Rh-Hr types = $2(\overline{rh'rh'} + \overline{Rh_1Rh_1} + \overline{Rh_ZRh_1})(\overline{rh} + \overline{rh''} + \overline{Rh_0} + \overline{Rh_2}) = 2(0.20)(0.30)$ or approximately 12 per cent. For all three systems combined = $1 - (1 - 0.04)(1 - 0.12)(1 - 0.12) = 0.258$ or approximately 26 per cent.

Situations not infrequently arise where a man falsely accused of paternity wishes to have blood tests performed on himself and the child, but the mother of the child is not available for one reason or another. In such cases, the same principles apply as in cases of disputed maternity; that is, the chances of excluding paternity are approximately 26 per cent, or about half of the chances when all three of the parties are available.

More Complex Problems

At times, a married man correctly suspects that he is not the father of more than one of his wife's children. In such cases, the blood tests may exclude paternity for one or more of the children. Obviously, the chances that paternity will be excluded for at least one of the children will be greater than 50 per cent. Such evidence may be and has been used as grounds for divorce. If monovular twins are involved, the chances are not increased, because such twins are invariably of identical blood groups.

In immigration cases, both of the putative parents may be excluded in cases of fraudulent applications. The chances that the falsity of the claim will be disclosed by excluding at least one of the parents will necessarily be considerably greater than 50 per cent.

An unusual but interesting problem sometimes arises, when it is suspected that two newborn babies have been interchanged in a hospital. In such cases the blood tests are of considerable aid, because with two sets of parents and two children available the chances of obtaining conclusive results are very high. For example, the chances of solving a problem of interchange of babies for a single blood factor has been shown by Wiener[231] to be as follows:

$$V_D = 2\overline{\text{D}}\ \overline{\text{R}}^2 \tag{54}$$

where V_D is the chances of solving the problem, $\overline{\text{D}}$ is the frequency of the blood factor in population, and $\overline{\text{R}}$ is the frequency of persons lacking the factor. The maximun chance for a single blood factor is as high as 29.03 per cent, and occurs when the frequency of phenotype D is equal to 0.33 and $\overline{\text{R}}$ equals 0.67.

Wiener has estimated the chances of detecting interchange of infants by the A-B-O groups to be approximately 40 per cent, and by the M-N types also approximately 40 per cent, so that the combined chances for the two blood group systems would be approximately 64 per cent. If, in addition, the Rh-Hr tests are done, about 90 per cent of these problems can be solved.

Circumstantial Evidence of Paternity

As has already been pointed out, the blood tests cannot be used to prove paternity, only to exclude it, because of the possibility of coincidence of the blood types of a falsely accused man and the actual father. There are rare circumstances, however, when both the accused man and the child have a blood factor of very low incidence in the population, or both have a rare combination of blood factors. Such evidence, while of little testimonial value, may help to convince the accused man of his paternity, so that he may then be willing to accept the child as his own. As examples may be mentioned the occurrence in both the putative father and the child of the rare types rh′ or rh″. Other infrequent types sometimes encountered in both the putative father and child are Rh_1^w and K.

On the other hand, the chances of excluding paternity in cases where a false accusation has been made may be enhanced by the use of such factors as **hr** and $\mathbf{rh}^w$. For example, a man of type Rh_1Rh_2 cannot have a child of type rh or Rh_0 unless he is the carrier of the genes r or R^0 (genotypes R^zr, R^zR^0 or R^0r^y). If tests on his red cells with anti-**hr** serum give a negative reaction, then he cannot be a carrier of either of these genes and paternity is excluded. Moreover, since factor $\mathbf{rh}^w$ is always associated with factor **rh′** in the same agglutinogen, an individual of type Rh_1^wrh cannot have a child of type Rh_1Rh_1, nor can an individual of type Rh_1Rh_1 have a child of type Rh_1^wrh.

It should be mentioned that the existence of "super" Rh genes constitute a pitfall in the application of the Rh-Hr types in disputed paternity cases. For example, a putative father belonging to type Rh_1Rh_1 while the child belongs to type Rh_0, or the reverse, is not excluded if the man and child are both carriers of gene $\overline{R}^0$. Fortunately gene $\overline{R}^0$ betrays its presence by the high agglutinability of the red cells by anti-$\mathbf{Rh_0}$ serums, as has already been explained (cf. page 87).

Case Reports

The application of blood grouping tests in actual practice is best illustrated by citing a few cases (cf. Wiener,[232] Alvarez,[232a] Unger,[250] Sussman and Schatkin,[251,253] Witebsky and Wyelegala[254]).

Case 1. This case, which was tried before the Court of Special Sessions in New York City, is unusual because a man was accused of the paternity of *six* children born out of wedlock. When he denied the charge, blood grouping tests were carried out, upon order of the court, on the defendant and on the complainant and her six children, with the results shown in table 49. As can be seen, the complainant belongs to group O, three of her children to subgroup A_1, and three to subgroup A_2. The defendant belongs to subgroup A_1, corresponding to which there are three possible genotypes, according to the genetic theory. If we assume that the defendant is of genotype $I^{A_1}I^{A_2}$, half his children with a group O mother would be expected to be subgroup A_1 and half subgroup A_2, as actually occurred in this case.

As far as the M-N types are concerned, the complainant is type N and the defendant type M, and all six children are type MN, as would necessarily result, assuming the defendant to be their father. Moreover, the complainant is type Rh_1rh, and the defendant type Rh_1Rh_1, three of the children type Rh_1-Rh_1, and three of type Rh_1rh, again in agreement with the theoretic expectations.

The excellent fit of the blood group findings of the six children with the genetic expectations strongly suggests that the defendant actually is their father, as charged. While blood tests that fail to exclude paternity are inconclusive because of the possibility of coincidence, the results here are so striking that they helped convince the defendant of his paternity, so that he contributed willingly to the children's support, in conformity with the decision of the court.

The use here of tests for the subgroups of A is inconsistent with the opinion already cited that these tests are not sufficiently reliable for routine use in medicolegal cases. However, subgrouping tests are so easy to perform and the results so informative that the senior author carries them out regularly. At times, the findings are illuminating, as in the present case. When tests for the subgroups of A indicate an exclusion of paternity but all other tests are inconclusive, the safest practice is to omit the subgrouping results from the official report, and to present them instead in a covering letter, explaining in detail the limitations of the test, for the guidance of the court.

Case 2. A man accused of the paternity of a child denied the charge. As shown in table 50, the blood tests indeed showed that he was not the father of the child. In fact, the results in this case were unusual in that the tests provided triple proof of nonpaternity.

Case 3. Four young Chinese adults came to the United States claiming derivative American citizenship. Upon the request of the Bureau of Immigration, blood grouping tests were carried out on these four individuals and their supposed parents. Since, as shown in table 51, the supposed mother belongs to group A_1B, the first and fourth applicants, both of whom belong to group O, cannot be her children. Moreover, since the two supposed parents are to types Rh_1Rh_2 and Rh_1rh, respectively, the second and third children who belong to types Rh_1Rh_1 and Rh_ZRh_1, respectively, cannot both be children of the couple.

TABLE 49

RESULTS OF BLOOD GROUPING TESTS IN A CASE OF DISPUTED PATERNITY INVOLVING SIX CHILDREN

Blood Group System		Putative Father	Mother	Children					
				1st	2nd	3rd	4th	5th	6th
A-B-O	*Phentotypes*	A_1	O	A_2	A_2	A_2	A_1	A_1	A_1
	Genotypes	$I^{A_1}I^{A_1}$, $I^{A_1}I^O$, or $I^{A_1}I^{A_2}$	I^OI^O	$I^{A_2}I^O$	$I^{A_2}I^O$	$I^{A_2}I^O$	$I^{A_1}I^O$	$I^{A_1}I^O$	$I^{A_1}I^O$
M-N	*Phenotypes*	M	N	MN	MN	MN	MN	MN	MN
	Genotypes	L^ML^M	L^NL^N	L^ML^N	L^ML^N	L^ML^N	L^ML^N	L^ML^N	L^ML^N
Rh-Hr	*Phenotypes*	Rh_1Rh_1	Rh_1rh	Rh_1rh	Rh_1Rh_1	Rh_1Rh_1	Rh_1rh	Rh_1rh	Rh_1Rh_1
	Genotypes	R^1R^1 or R^1r'	R^1r, R^1R^0, or R^0r'	R^1r, R^1R^0, or R^0r'	R^1R^1 or R^1r'	R^1R^1 or R^1r'	R^1r, R^1R^0, or R^0r'	R^1r, R^1R^0, or R^0r'	R^1R^1 or R^1r'

Thus, the claims of derivative citizenship of at least three of the four applicants were proved to be fraudulent.

Nomenclature

The current controversy regarding Rh-Hr nomenclature has already been discussed in detail (cf. page 100). This important question has been the subject of study by a special Subcommittee on Blood Grouping Tests, of the Committee on Medicolegal Problems of the American Medical Association. As is pointed out in their report,[233] "The submission to jurists by different experts of medicolegal reports with conflicting symbolisms cannot help but confuse them and shake their confidence in the blood tests. Moreover, reports in two different symbolisms require translation from one into the other when comparisons are to be made; there is always danger of error in the process. Therefore, the adoption of one uniform nomenclature for medicolegal reports is most desirable."

From a study of the available evidence, the committee concluded, in part, as follows: "Although corresponding to the two systems of designating the Rh-Hr types there are two conflicting genetic theories, the crux of the problem is actually on a serologic level. It is most important to distinguish clearly between agglutinogens and their serologic attributes, the blood factors . . . . The C-D-E notations for the Rh-Hr types make no allowance for the difference between a blood factor and an agglutinogen and incorporate the tacit, incorrect assumption that every agglutinogen has but a single corresponding antibody. This has led to a number of paradoxes . . . . Indeed much of the evidence that has accumulated regarding the serology and genetics of the Rh-Hr types refutes the interpretation implicit in the C-D-E notations. On the other hand, the original Rh-Hr nomenclature presents the data objectively, without committing the user to a preconceived interpretation . . . . The committee, therefore, recommends that the C-D-E notations for the Rh-Hr types be discarded and that, in approved medicolegal reports, unless and until some other convention is agreed upon the original Rh-Hr nomenclature be retained as the sole nomenclature for this blood group system."

Recently, the recommendation that the Rh-Hr nomenclature be used exclusively for medicolegal reports has been approved by the Section on Immunology of the American Academy of Forensic Sciences. The time has now come to extend this recommendation to the applications of the Rh-Hr types in clinical medicine and anthropology. Confusion will remain until editors of scientific journals require that all articles dealing with the Rh-Hr types conform with this recommendation, as has been done in the case of the analogous recommendation made in a report[234] published in 1937, regarding the nomenclature of the A-B-O blood groups.

Qualifications of Experts

The field of immunohematology has grown in scope and complexity, and there are relatively few workers who have devoted the time and effort required to become expert in the field. As a guide for lawyers and courts desiring to use blood tests as an aid in cases of disputed parentage, the Law Department of the American Medical Association has available a list of qualified blood group specialists. A similar panel of qualified workers is used by the Court of Special Sessions in New York City, from which lawyers may select an expert to carry out blood tests in cases of disputed paternity. However, failure of many jurisdictions to exert equal care in the selection of experts to carry out blood grouping examinations has been responsible for a number of unfortunate mistakes, and this is a problem which still requires a satisfactory solution.

In blood transfusion practice, it is considered *a priori* evidence of negligence if a patient of group O is given group A blood, and if the patient dies, the physician, hospital and/or blood bank may be subjected to costly lawsuits. Similarly, a mistake in blood grouping

TABLE 50

TRIPLE EXCLUSION BY BLOOD TESTS OF MAN FALSELY ACCUSED OF PATERNITY

Blood Group System		Blood of		
		Putative father	Mother	Child
A-B-O	*Phenotypes*	A_2	O	B
	Genotypes	$I^{A_2}I^{A_2}$ or $I^{A_2}I^O$	I^OI^O	I^BI^O
M-N	*Phenotypes*	N	N	MN
	Genotypes	L^NL^N	L^NL^N	L^ML^N
Rh-Hr	*Phenotypes*	Rh_1Rh_1	rh″	rh
	Genotypes	R^1R^1 or R^1r'	$r''r$	rr

TABLE 51

CASE ILLUSTRATING THE APPLICATION OF BLOOD TESTS TO DISPROVE FALSE CLAIMS OF DERIVATIVE CITIZENSHIP

Blood Group System		Putative Father	Putative Mother	Children			
				1st	2nd	3rd	4th
A-B-O	*Phenotypes*	A_1	A_1B	O	A_1B	A_1	O
	Genotypes	$I^{A_1}I^{A_1}$, $I^{A_1}I^{A_2}$, or $I^{A_1}I^O$	$I^{A_1}I^B$	I^OI^O	$I^{A_1}I^B$	$I^{A_1}I^{A_1}$, $I^{A_1}I^{A_2}$ or $I^{A_1}I^O$	I^OI^O
M-N	*Phenotypes*	MN	M	MN	M	MN	M
	Genotypes	L^ML^N	L^ML^M	L^ML^N	L^ML^M	L^ML^N	L^ML^M
Rh-Hr	*Phenotypes*	Rh_1Rh_2	Rh_1rh	Rh_1Rh_1	Rh_1Rh_1	Rh_ZRh_1	Rh_1rh
	Genotypes	R^1R^2, R^1r'', R^2r', R^zr, R^zR^0, R^0r^y, R^zR^2, R^zr'', or R^2r^y	R^1r, R^1R^0, or R^0r'	R^1R^1, or R^1r'	R^1R^1, or R^1r'	R^zR^1, R^zr', or R^1r^y	R^1r, R^1R^0, or R^0r'

in a medicolegal case of disputed paternity erroneously excluding paternity or maternity could conceivable lead to a lawsuit for malpractice or for damages for defamation of character. An error in medicolegal cases is even less defensible than in clinical cases, because the claim that the tests had to be done under the pressure of an emergency cannot be used. The medicolegal expert has plenty of time in which to carry out the tests, as many times as desired and with as many controls as necessary. When in doubt, he can draw new samples and repeat the tests, and if he wishes he can send the blood samples to an independent expert for confirmation. Unfortunately, however, the least experienced workers, whose reports are most in need of verification, are generally the least sensitive to the possibility that their findings may be in error.

A fully qualified expert in blood grouping is not merely a careful technician, but an individual thoroughly versed in the fundamentals of blood group serology and genetics. Such an individual prefers accuracy to simplicity, and has no difficulty in using the recommended Rh-Hr nomenclature, which merely translates the results obtained into symbols. Such a worker has no difficulty in preparing a clear self-explanatory report, when presenting his evidence to a lay bench and jury. Wiener[235] maintains, therefore, that with certain outstanding exceptions, those who find it necessary to resort to the incorrect C-D-E notations for the sake of "simplicity" do not understand the subject thoroughly, and he suggests that this may be used as one of the criteria by which to judge the qualifications of an expert.

CHAPTER XII

Anthropological Studies on the Blood Types

Until recently, the use of blood grouping tests in racial studies attracted relatively little attention among physical anthropologists. Thus, most journals and text books on physical anthropology concentrate a good deal of attention on skeletal measurements and external physical characteristics, but devote little space to blood grouping. This has been due principally to the limited value of the findings until the Rh-Hr blood types were discovered. For example, when tests for the A-B-O groups alone are carried out, closely related ethnic groups may exhibit significant differences in distribution while totally unrelated peoples may have similar distributions. The situation was improved somewhat by the introduction of the subgroups of A, and the M-N types, but the greatest advances resulted from the discovery of the Rh-Hr types.

At first the studies were restricted to the $\mathbf{Rh_0}$ factor alone, because the only serums available for the earlier studies were the animal antirhesus serums, or human serums of specificity anti-$\mathbf{Rh_0}$. It was soon observed that while the $\mathbf{Rh_0}$ factor is present among approximately 85 per cent of Caucasoids, it occurs more frequently among Negroids, and among practically 100 per cent of Mongoloids.[236] The results were quite striking, and became more so when the tests were extended to include the **rh′** and **rh″** factors. For example, Negroids were characterized by a high frequency of type Rh_0, a type of blood which occurs infrequently among Caucasoids and Mongoloids.[237] The introduction of tests for the factors **hr′** and **hr″** did not alter the situation to any great degree, because, as has already been pointed out, the reactions given by anti-Hr serums can largely be predicted in advance from the reactions of the anti-Rh serums. The principal value of Hr serums in anthropological investigations is as an aid in recognizing individuals carrying the rare gene R^z. In this way it was possible to demonstrate that gene R^z, which is rare among Caucasoids, and has not been demonstrated to date among Negroids, occurs in as many as 3 to 6 per cent of Mongoloids.[238]

Based on the newer findings concerning the distribution of the blood group factors, especially Rh-Hr, Wiener has proposed a semi-qualitative classification of human races as follows:

I. Caucasoid group—highest frequency of gene r; gene I^{A_2} present as well as gene I^{A_1}; gene L^M slightly more frequent than gene L^N.

II. Negroid group—highest frequency of gene R^0 and Rh_0 variants; gene I^{A_2} present as well as gene I^{A_1}, also frequent A intermediates; gene L^M slightly more frequent than gene L^N.

III. Mongoloid group—highest frequency of the rare gene R^z; genes r and I^{A_2} absent or rare.

   a. Asiatic subdivision—genes L^M and L^N almost equal in frequency.

   b. Pacific subdivision (including Australian aborigines and Ainu)—low frequency of gene L^M and high frequency of gene L^N.

   c. American subdivision (including Amerindians and Eskimos)—high frequency of gene L^M and low frequency of gene L^N.

Before the use of blood tests was introduced, anthropologists classified races on the basis of their external physical characteristics or skeletal measurements. This led to a number of manifest contradictions and paradoxes. One reason for this is that the external physical characteristics are subject to conscious selection while conscious selection would hardly play a role for the blood groups. Moreover, whereas physical characteristics such as skin color, texture of the hair, height, and shape of the nose have considerable adaptive value, the possession or lack of any particular blood factor has no special advantage, except in relation to isosensitization in pregnancy. Based on their physical characteristics Australian aborigines were considered to be intermediate between Negroids and Caucasoids, while Papuans were placed in the Negroid group. Studies on the Rh-Hr types, on the other hand, clearly demonstrated that these peoples belong to the Mongoloid group, a conclusion which is more reasonable judging from geographical and historical considerations.

Study of the distribution of the blood group factors, especially the Rh-Hr types, is of great value in analyzing the results of crosses between different ethnic groups. For example, the distribution of the Rh types among Puerto Ricans (cf. table 25) clearly indicates that this ethnic group arose from a cross between Negroids and Caucasoids. However, one should be cautious before drawing conclusions from the results of blood tests alone. For example, in the case of Puerto Ricans the presence of shovel-shaped incisors indicates that Mongoloids, that is, American Indians, also participated in the cross from which the Puerto Ricans originated, as had been pointed out by Washburn.

That the A-B-O groups and Rh-Hr types are not entirely without selective value has been demonstrated by the discovery of their role in isosensitization by pregnancy. For example, Rh-negative mothers (genotype *rhrh*) who have become sensitized to the $\mathbf{Rh_0}$ factor will have a certain number of Rh-positive babies (genotype *Rhrh*) who are stillborn or die during the neonatal period from erythroblastosis. As Wiener[239] pointed out, if we assume that we are dealing with a population of constant size containing x *Rh* genes and y *rh* genes, then the initial distribution of the genes would be as follows:

$$Rh = \frac{x}{x + y}; \text{ and } rh = \frac{y}{x + y}$$

If the number of fetuses and newborn that die from erythroblastosis during one generation is c, then the distribution of the genes during the second generation would be:

$$Rh = \frac{x - c}{x + y - 2c}; \text{ and } rh = \frac{y - c}{x + y - 2c}$$

If to begin with the number of *Rh* genes is equal to the number of *rh* genes, this process would have no effect on the relative distribution of the genes. If, on the other hand, the incidence of the two genes is initially unequal, the less frequent gene would be affected to a greater extent than the more common gene, so that eventually, other things being equal, over a period of thousands of generations, the incidence of the less common gene would be substantially reduced or it might be practically eliminated.

This process could explain the virtual absence of the $\mathbf{Rh_0}$-negative type among Mongoloids, but it then becomes difficult to account for the relatively high frequency of gene *rh* among Caucasoids. Wiener[239] and Haldane[240] independently pointed out that if one assumed the existence of a population in the past, and possibly still existing at the present time, consisting largely of Rh-negative persons, a hybrid population could result with a serological composition resembling that of Caucasoids. This hypothesis has received support from the findings in Basques by Etcheverry,[241] and confirmed by Mourant. Among these people, who differ from other Europeans in their isolated habitat, unique language and social customs, the frequency of gene *rh* is extraordinarily high, namely, about 60 per cent. It had been previously shown by the Hirszfelds[242] that the incidence of gene I^B decreases from east to west in Eurasia. According to Candela[243] gene I^B was introduced at the time of the Mongolian invasion, the degree of crossing between European and Mongoloids being indicated by the frequency of gene I^B. Here again the Basques are unique in that the frequency of gene I^B is the lowest in Europe. For these reasons it seems probably that the Basques represent most closely the original European stock which must have had a very high frequency of gene *rh* as well as a very low frequency of gene I^B.

Analysis of Mixed Populations

A number of methods have been suggested for analyzing blood group frequencies of mixed populations. The principle of the calculations is the same as for computing the results of mixing solutions of different concentrations. Here we shall describe a simple method which has been used by Mourant.[244] This method has the advantage that it can be applied to many blood factors at the same time, and thus give a comprehensive view.

Let p represent the frequency of the gene for a given blood factor in a population A; and let q represent the frequency of the same gene in a population B. Then, suppose a new population C if formed with x individuals from population A, and y individuals from population B. If r represents the frequency of the gene in the mixed population, this is obviously the weighted mean of p and q, and is readily calculated by the following formula:

$$r = \frac{px + qy}{x + y} \tag{55}$$

The value of r can also be derived geometrically. The frequency p for population A is erected on one vertical line, and the frequency q for population B on another vertical line. The interval between the two vertical lines, A and B, is divided in inverse proportion to x and y, and a new vertical line, C, erected. A straight line is then drawn connecting the points p and q, and the point at which it intersects line C represents the frequency of the gene in the mixed population.

Conversely, if it is known that a certain mixed population, C, resulted from a cross between two other populations, A and B, the proportion of individuals from populations A and B which entered the cross can be determined if the frequencies of the gene for any agglutinogen is known for all three populations. One first marks the frequency p for population A on one vertical line, and the frequency q for population B on a second vertical line, and a straight line is drawn between these two points. A second line is then drawn parallel with the horizontal axis at distance r from it. At the point of intersection of the two points a third line is erected, and this represents the composition of the population C.

An example of the practical application of the method is given in figure 4, from the study of Chalmers et al.[245] In constructing this diagram the frequencies of gene r were used as the basis. A straight line was drawn connecting the points representing the frequencies of gene r in Basques on the one hand, and Siamese on the other. Then the horizontal lines were drawn representing the frequencies of gene r in England, Latvia and India, and through points of intersection obtained, vertical lines were then erected for these three populations. On the five vertical lines points representing the fre-

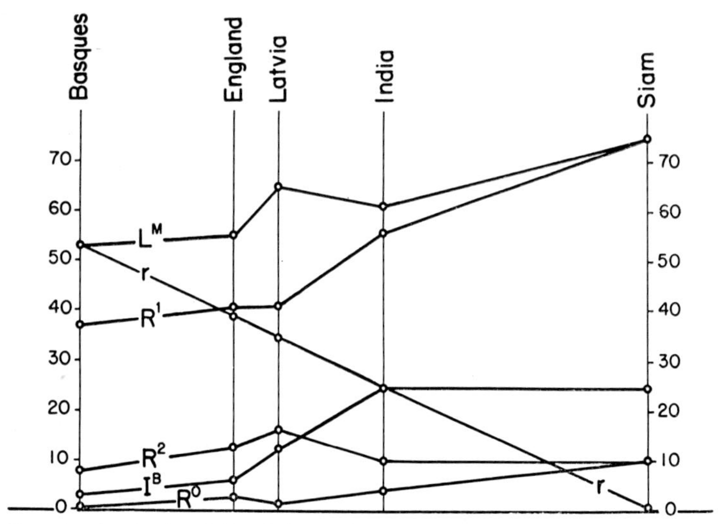

Fig. 4.—Analysis of Blood Group Gene Frequencies of Mixed Populations. (After Chalmers et al.)

quencies of genes R^1, R^2, R^0, I^B, and L^M were marked and the corresponding points on the five lines connected producing the results shown in the figure. If the populations in England, Latvia and India had been produced as a result of crossing between Basques and Siamese one would expect the lines representing each of the gene frequencies to be straight. As a matter of fact, the lines are not all straight which is to be expected since the populations in question are not merely such simple mixtures.

The purpose of this chapter has been to present the broad principles involved in the application of blood groups in anthropology. No attempt has been made to tabulate the numerous studies that have been reported. Readers who have occasion to consult the original sources should bear in mind that some of the data in the literature suffer from errors in blood typing, and this possibility should always be borne in mind especially when the findings appear to be bizarre or unique. The first compendium of the literature was prepared by Boyd[246] and included principally the earlier studies on the A-B-O groups. In Wiener's book[247] a less complete compilation was given, but this source also lists findings on the subgroups of A and the M-N types. The most comprehensive and most valuable source is the recent book by Mourant,[248] which includes the voluminous data especially on the Rh-Hr blood types, and also on other blood group systems, including P, Kell, Duffy, Kidd, and Lutheran. One of the most important investigators in this field is R. T. Simmons,[258–260] whose numerous valuable contributions should be studied and used as models by those interested in the subject.

References

1. a. Landsteiner, K.: *The Specificity of Serological Reactions*, 2nd rev. ed., 1945, Harvard University Press, Cambridge, Mass.
   b. Wiener, A. S.: *Blood Groups and Transfusion*. 3rd ed., 1943, C. C. Thomas, Springfield, Ill.
   c. Wiener, A. S.: *Rh-Hr Blood Types, Applications in Clinical and Legal Medicine and Anthropology*. 1954, Grune & Stratton, New York.
   d. Wiener, A. S.: *Rh-Hr Syllabus*. 1954, Grune & Stratton, New York; 1955, Georg Thieme, Stuttgart; 1955, V. Idelson, Naples; 1956, Prensa Med. Mexicana, Mexico; 1957, Tokyo (Japanese).
   e. Race, R. R., and Sanger, R.: *Blood Groups in Man*. 3rd ed., 1958, Blackwell Scientific Publ., Oxford.
   f. Schiff, F., and Boyd, W. C.: *Blood Grouping Technic*. 1942, Interscience Publishers, Inc., New York, N. Y.
2. Landsteiner, K.: *Zentralbl. f. Bakteriol.*, **27:** 357 (1900); *Wien. klin. Woch.*, **14:** 1132 (1901).
3. Landsteiner, K., and Levine, P.: *Proc. Soc. Exp. Biol. and Med.*, **24:** 600, 941 (1927).
4. Landsteiner, K., and Wiener, A. S.: *Proc. Soc. Exp. Biol. and Med.*, **43:** 223 (1940).
5. Wiener, A. S., and Sonn-Gordon, E. B.: *Amer. J. Clin. Path.*, **17:** 67 (1947).
6. a. Renkonen, K. O.: *Ann. Med. Exp. Biol., Fenn.*, **26:** 66 (1948).
   b. Krüpe, M.: *Blutgruppenspezifische Pflanzliche Eiweisskörper*, 1956, Ferdinand Enke, Stuttgart.
7. Fisher, R. A.: *Statistical Methods for Research Workers*, 6th ed., 1936, Oliver & Boyd, Edinburgh.
8. a. Wiener, A. S., and Wexler, I. B.: *Bact. Rev.*, **16:** 69 (1952).
   b. Wiener, A. S., and Wexler, I. B.: L. Gedda's *Novant' Anni delle Leggi Mendeliane*, 1956, pp. 147–162, Istituto Gregorio Mendel, Rome.
   c. Wiener, A. S., and Karowe, H.: *J. Immunol.*, **49:** 51 (1944).
9. Wiener, A. S.: *J. Immunol.*, **34:** 87 (1938).
10. Epstein, A. A., and Ottenberg, R.: *Trans. N. Y. Pathol. Soc.*, **8:** 187 (1908).
11. von Dungern, E., and Hirszfeld, L.: *Ztschr f. Immunitäts.*, **6:** 284 (1910).
12. Bernstein, F.: *Ztschr. f. indukt. Abstamm. u. Vererbungslehre*, **37:** 237 (1925).
13. Kirihara, S., and Hakurinsai, K.: *Nagoya J. Med. Sci.*, **2:** 75 (1932).
14. Furuhata, T.: *Japan Med. World*, **7:** 197 (1927).
15. Landsteiner, K., and Levine, P.: *Proc. Soc. Exp. Biol. and Med.*, **24:** 941 (1927).
16. Thomsen, O., Friedenreich, V., and Worsaae, E.: *Acta Path. et Microbiol. Scand.*, **7:** 157 (1930).

17. Landsteiner, K., and Levine, P.: *J. Exp. Med.*, **48:** 731 (1928).
18. Sanger, R., and Race, R. R.: *Nature* (Lond.), **160:** 505 (1947).
19. Walsh, R. J., and Montgomery, C.: *Nature* (Lond.), **160:** 504 (1947).
20. Wiener, A. S.: *Acta Genet. Med. Gemellol.*, **3:** 314 (1954).
21. Wiener, A. S., Unger, L. J., and Gordon, E. B.: *J. Amer. Med. Assoc.*, **153:** 1444 (1953).
22. Greenwalt, T., Sasaki, T., Sanger, R., and Race, R. R.: *Proc. Nat. Acad. Sci.*, **40:** 1126 (1956).
23. Shapiro, M.: *J. Forensic Med.* (S. Africa), **3:** 152 (1956).
24. Landsteiner, K., and Wiener, A. S.: *J. Exp. Med.*, **74:** 309 (1941).
25. Wiener, A. S.: *Science*, **98:** 182 (1943).
26. Wiener, A. S.: *Proc. Soc. Exp. Biol. and Med.*, **54:** 316 (1943).
27. Murray, J., Race, R. R. and Taylor, G. L.: *Nature* (Lond.), **153:** 560 (1944).
28. Wiener, A. S.: *Nature* (Lond.) **2:** 735 (1948).
29. Wiener, A. S.: *Science*, **100:** 595 (1944).
30. Stormont, C., Owen, R. D., and Irwin, M. R.,: *Genetics*, **36:** 134 (1951).
31. Sanger, R.: *Nature* (Lond.) **176:** 1163 (1955).
32. Levine, P., Bobbitt, O. B., Waller, W. K., and Kuhmichel, A. A.: *Proc. Soc. Exp. Biol. and Med.*, **77:** 403 (1951).
33. Allen, F. H., and Lewis, S. J.: *Vox Sang.*, **7:** 8 (1957).
34. a. Li, C. C.: *Population Genetics*, 1955, Univ. Chicago Press, Chicago.
b. Kempthorne, O.: *An Introduction to Genetic Statistics*, 1957, John Wiley & Sons, Inc., N. Y.
35. Hirszfeld, L., and Hirszfeld, H.: *Lancet*, **2:** 675 (1919).
36. Bernstein, F.: *Ztschr. f. indukt. Abstamm. u. Verebungsl.*, **56:** 233 (1930).
37. Wiener, A. S.: *Science*, **100:** 595 (1944).
38. Henry, N. R.: *Med. J. Austral.*, **1:** 395 (1946).
39. Schiff, F.: *Technik der Blutgruppenuntersuchung*, 1932, Julius Springer, Berlin.
40. Landsteiner, K., and Levine, P.: *J. Immunol.*, **12:** 441 (1926).
41. Fischer, W., and Hahn, F.: *Ztschr. f. Immunitäts.*, **84:** 177 (1935).
42. Wiener, A. S., and Silverman, I. J.: *Amer. J. Clin. Path.*, **11:** 45 (1941).
43. Friedenreich, V.: *Ztschr. f. Immunitäts.*, **89:** 409 (1936).
44. Wiener, A. S.: *Ann. Eugen.* (Lond.) **18:** 1 (1953).
45. Gammelgaard, A., and Marcussen, A.: *Ztschr. f. Immunitäts.*, **98:** 411 (1940).
46. Grove-Rasmussen, M., Soutter, L., and Levine, P.: *Amer. J. Clin Path.*, **22:** 1157 (1952).
47. Dunsford, I., and Aspinall, P.: *Ann. Eugen.* (Lond.), **17:** 30 (1952).
48. Unger, L. J., and Wiener, A. S.: J. Lab. & Clin. Med. **44:** 387 (1954).
49. Schiff, F., and Adelsberger, L.: *Ztschr. Immunitäts.*, **40:** 335 (1924).
50. Friedenreich, V., and With, S.: *Ztschr. Immunitäts.*, **78:** 152 (1933).
51. Owen, R. D.: *J. Immunol.*, **73:** 29 (1954).
52. Lehrs, H.: *Ztschr. Immunitäts.*, **66:** 175 (1930).
53. Putkonen, T.: *Acta Med. Fenn.* "Duodecim" A 14, Part 2 (1930).
54. Wiener, A. S., and Belkin, R. B.: *J. Immunol.*, **47:** 467 (1943).

55. Schiff, F., and Sasaki,: *Ztschr. Immunitäts.*, **77:** 129 (1932); *Klin. Woch.*, **34:** 1426 (1932).
56. Uyeyama, R.: *Jap. J. Med. Sci.*, **3:** 13 (1940).
57. Mourant, A. E.: *Nature* (Lond.) **158:** 237 (1946).
58. Andresen, P. H.: *Acta Path. et Microbiol. Scand.*, **24:** 616 (1948).
59. Andresen, P. H.: *Acta Path. et Microbiol. Scand.*, **25:** 728 (1948).
60. Grubb, R.: *Nature* (Lond.) **162:** 933 (1948).
61. Morgan, W. T. J.: *Proceedings Vth International Congress of Blood Transfusion*, pp. 285–299, 1955; Annison, E. F., and Morgan, W. T. J.: *Biochem. J.* **50:** 462 (1952).
62. Race, R. R., Sanger, R., Lawler, S. D., and Bertinshaw, D.: *Brit. J. Exp. Path.*, **30:** 73 (1949).
63. Andresen, P. H., Andresen, A., Jordal, K., and Henningsen, L.: *Rev. d'Hématol.*, **5:** 305 (1950).
64. Sneath, J. S., and Sneath, P. H. A.: *Nature* (Lond.) **176:** 172 (1955).
65. Nicholas, J. W., Jenkins, W. L., and Marsh, W. L.: *Brit. Med. J.* **1:** 1458 (1957).
66. Grubb, R., and Morgan, W. T. J.: *Brit. J. Exp. Path.* **30:** 198 (1949).
67. Wiener, A. S.: *Lab Digest* (St. Louis) Vol. 18, No. 5, November, 1954.
68. Cepellini, R.: personal communication.
69. Landsteiner, K., and Miller C. P.: *J. Exp. Med.*, **43:** 835 (1925).
70. Wiener, A. S.: unpublished observations.
71. Wiener, A. S., Candela, P. B., and Goss, L. J.: *J. Immunol.*, **45:** 229 (1942).
72. Candela, P. B., Wiener, A. S. and Goss, L. J.: *Zoölogica*, **25:** 513 (1940).
73. Wiener, A. S., and Gordon, E. B.: *Brit. J. Hematol.*, **2:** 305 (1956).
74. Weiner, W., Lewis, B., Moores, P., Sanger, R., and Race, R. R.: *Vox Sang.* **2:** 25 (1957).
75. Yokoyama, M., Stacey, S. M., and Dunsford, I.: *Vox Sang.* **2:** 348 (1957).
76. Bhende, Y. M., Deshpande, C. K., Bhatia, H. M., Sanger, R., Race, R. R., Morgan, W. T. J., and Watkins, W.: *Lancet* **1:** 903 (1952).
77. Levine, P., Robinson, E., Celano, M., Briggs, O., and Falkinburg, L.: *Blood*, **10:** 1100 (1955).
78. Landsteiner, K., and Levine, P.: *J. Exp. Med.*, **47:** 757 (1928).
79. Wiener, A. S., Zinsher, R., and Selkowe, J.: *J. Immunol.*, **27:** 431 (1934).
80. Landsteiner, K., and Levine, P.: *J. Exp. Med.*, **48:** 731 (1928).
81. Wiener, A. S.: *J. Immunol.*, **21:** 157 (1931); *Human Biology*, **7:** 222 (1935).
82. Crome, W.: *Deutsche Ztschr. f. d. ges. gerichtl. Med.*, **24:** 167 (1935).
83. Friedenreich, V.: *Deutsche Ztschr. f. d. ges. gerichtl. Med.*, **25:** 358 (1936).
84. Wiener, A. S.: *J. Immunol.*, **34:** 87 (1938).
85. Landsteiner, K., and Wiener, A. S.: *J. Immunol.*, **33:** 19 (1937).
86. Walsh, R. J., and Montgomery, C.: *Nature* (Lond.) **160:** 504 (1947).
87. Sanger, R., Race, R. R., Walsh, R. J., and Montgomery, C.: *Heredity*, **2:** 131 (1948).
88. Wiener, A. S., DiDiego, N., and Sokol, S.: *Acta Med. Genet. et Gemellol.*, **2;** 391 (1953).

89. Wiener, A. S.: *Amer. J. Human Genet.*, **4:** 463 (1952).
90. Levine, P., Kuhmichel, A. B., Wigod, M., and Koch, E.: *Proc. Soc. Exp. Biol. and Med.*, **78:** 218 (1951).
91. Wiener, A. S.: *Acta Med. Genet. et Gemellol.*, **3:** 314 (1954).
92. Wiener, A. S.: *Acta Med. Genet. et Gemellol.*, **6:** 95 (1957).
93. Fisher, R. A.: Cited after Race, R. R., and Sanger, R.: *Blood Groups in Man*, 2nd ed., p. 62, 1954.
94. Race, R. R., and Sanger, R.: *Blood Groups in Man*, 2nd ed., 1954.
95. Wiener, A. S., Unger, L. J., and Gordon, E. B.: *J. Amer. Med. Assoc.*, **153:** 1444 (1953).
96. Wiener, A. S., Unger, L. J., and Cohen, L.: *Science*, **119:** 734 (1954).
97. Greenwalt, T., et al.: *Proc. Nat. Acad. Sci.*, **40:** 1126 (1956).
98. Landsteiner, K., Strutton, W. R., and Chase, M. W.: *J. Immunol.*, **27:** 469 (1934).
99. Chalmers, J. N. M., Ikin, E. W., and Mourant, A. E.: *Brit. Med. J.*, **2:** 175 (1953).
100. Shapiro, M.: *J. Forensic Med.* (S. Afr.) **3:** 152 (1956).
101. van der Hart, M., Bossman, H., and van Loghem, J. J.: *Vox Sang.*, **4:** 108 (1954).
102. Levine, P., et al.: *Proc. Soc. Exp. Biol. and Med.*, **77:** 402 (1951).
103. Wallace, J., Milne, G. R., Mohn, J., Moores, P., Sanger, R., and Race, R. R.: *Nature* (London), **179:** 478 (1957).
104. Allen, R., Corcoran, P. A., Kenton, H. B., and Breare, N.: in press.
105. Landsteiner, K. and Levine, P., *J. Immunol.*, **17:** 1 (1929).
106. Wiener, A. S., and Peters, H. R.: *Ann. Int. Med.*, **13:** 2306 (1940); Wiener, A. S.: *Amer. J. Clin. Path.*, **12:** 302 (1942).
107. Furuhata, T., and Imamura, I.: *Jap. J. Genetics* (Japanese) **12:** 50 (1926).
108a. Wiener, A. S.: *J. Immunol.* **66:** 287 (1951).
108b. Cameron, G. L., and Staveley, J. M.: *Nature* (London) **179:** 147 (1957).
109. Levine, P., Bobbitt, O., Waller, R. K., and Kuhmichel, A.: *Proc. Soc. Exp. Biol. and Med.*, **77:** 403 (1951).
110. Sanger, R.: *Proc. 6th Cong. Internat. Soc. of Blood Transfusion*, pp. 110–113, S. Karger, N. Y., 1958.
111. Landsteiner, K., and Wiener, A. S.: *Proc. Soc. Exp. Biol. and Med.*, **43:** 223 (1940).
112. Wiener, A. S.: *Proc. Soc. Exp. Biol. and Med.*, **70:** 576 (1949).
113. Landsteiner, K., and Wiener, A. S.: *J. Exp. Med.*, **74:** 309 (1941).
114. Wiener, A. S.: *Arch. Path.*, **32:** 227 (1941).
115. Wiener, A. S.: *Proc. Soc. Exp. Biol. and Med.*, **56:** 173 (1944).
116. Race, R. R.: *Nature* (London), **153:** 771 (1944).
117. Diamond, L. K., and Abelson, N. M.: *J. Clin. Investig.*, **24:** 122 (1945).
118. Wiener, A. S.: *J. Lab. and Clin. Med.*, **30:** 662 (1945).
119. Diamond, L. K., and Denton, R. L.: *J. Lab. and Clin. Med.*, **31:** 821 (1945).
120. Coombs, R. R. A., Mourant, A. E., and Race, R. R.: *Nature* (London), **2:** 15 (1945).

121. MORTON, J. A., AND PICKLES, M. M.: *Nature* (London), **159:** 779 (1947).
122. WIENER, A. S., AND LANDSTEINER, K.: *Proc. Soc. Exp. Biol. and Med.*, **53:** 167 (1943).
123. WIENER, A. S.: *Science*, **98:** 182 (1943); Wiener, A. S. and Sonn, E. B.: *J. Immunol.*, **47:** 461 (1943).
124. WIENER, A. S.: *Proc. Soc. Exp. Biol. and Med.*, **54:** 316 (1943).
125. WIENER, A. S., SONN, E. B., AND BELKIN, R. B.: *Proc. Soc. Exp. Biol. and Med.*, **54:** 238 (1943).
126. WIENER, A. S., SONN, E. B., AND BELKIN, R. B.: *J. Exp. Med.*, **56:** 173 (1944).
127. WIENER, A. S., AND HYMAN, M. A.: *Amer. J. Clin. Path.*, **18:** 921 (1948).
128. WIENER, A. S.: *Nature* (London), **2:** 735 (1948).
129. Cited by LEVINE, P.: *J. Ped.*, **23:** 656 (1943).
130. RACE, R. R., AND TAYLOR, G. L.: *Nature* (London), **152:** 300 (1943).
131. WIENER, A. S., DAVIDSOHN, I., AND POTTER, E. L.: *J. Exp. Med.*, **81:** 63 (1945).
132. FISHER, R. A., cited after RACE, R. R.: *Nature* (London), **153:** 771 (1944).
133. MOURANT, A. E.,: *Nature* (London), **155:** 544 (1945).
134. WIENER, A. S., AND PETERS, H. R.: *Amer. J. Clin. Path.*, **18:** 533 (1948).
135. WIENER, A. S.: *Science*, **102:** 479 (1945).
136. WIENER, A. S., DAVIDSOHN, I., AND POTTER, E. L.: *J. Exp. Med.*, **81:** 63 (1945).
137. WIENER, A. S., SONN, E. B., AND POLIVKA, H.: *Proc. Soc. Exp. Biol. and Med.*, **61:** 382 (1946).
138. WIENER, A. S., AND SONN-GORDON, E. B.: *J. Immunol.*, **57:** 203 (1947).
139. WIENER, A. S.: Proc. Eighth Internat. Cong. Genetics, *Hereditas*, Suppl. Vol, pp. 500–519 (1949).
140. RACE, R. R., TAYLOR, G. L., IKIN, E. W., AND PRIOR, A. W.: *Ann. Eugen.* (London), **12:** 206 (1944).
141. RACE, R. R., TAYLOR, G. L., IKIN, E. W., AND DOBSON, A. M.: *Ann. Eugen.* (London), **12:** 261 (1945).
142. MCFARLANE, M.: *Ann. Eugen.* (London), **13:** 15 (1946).
143. LEVINE, P.: personal communication.
144. MCGEE, R. P., LEVINE, P., AND CELANO, M.: Science, **125:** 1043 (1957).
145. WIENER, A. S., AND BRANCATO, G. J.: *J. Lab. and Clin. Med.*, **40:** 27 (1952).
146. CALLENDER, S. T., AND RACE, R. R.: *Ann. Eugen.* (London), **13:** 102 (1946).
147. WIENER, A. S., BRANCATO, G. J., AND GORDON, E. B.: *Acta Genet. Med. et Gemellol.*, **6:** 195 (1957).
148. LAWLER, S. D., BERTINSHAW, D., SANGER, R., AND RACE, R. R.: *Ann. Eugen.* (London), **15:** 258 (1950).
149. WIENER, A. S., GORDON, E. B., AND HANDMAN, L.: *Amer. J. Human Genet.*, **1:** 127 (1949).
150. ROSENFIELD, R. E., VOGEL, P., GIBBEL, N., SANGER, R., AND RACE, R. R.: *Brit. Med. J.*, **1:** 975 (1953).
151. SANGER, R., RACE, R. R., ROSENFIELD, R. E., VOGEL, P., AND GIBBEL, N.: *Proc. Nat. Acad. Sci.*, **39:** 824 (1953).
152. WIENER, A. S.: *Brit. Med. J.*, **1:** 1391 (1953).

153. DeNatale, A., Cahan, A., Jack, J. A., Race, R. R., and Sanger, R.: *J. Amer. Med. Assoc.*, **159:** 247 (1955).
154. Wiener, A. S.: *Science*, **100:** 595 (1944).
155. Wiener, A. S., Unger, L. J., and Sonn, E. B.: *Proc. Soc. Exp. Biol. and Med.*, **58:** 89 (1945).
156. Rosenfield, R. E., Vogel, P., Miller, E. B., and Haber, G.: *Blood*, **6:** 1123 (1951).
157. Ceppellini, R., Dunn, L. C., and Turri, M.: *Genetics*, **41:** 283 (1955).
158. Levine, P., Celano, M., McGee, R., Muschel., L. H., and Griset, T. A.: *Hematological and Medico-Legal Reports*. (P. H. Andresen; papers in dedication of his 60th birthday), pp. 144–155, Munksgaard, Copenhagen (1957).
159. Race, R. R., Sanger, R., and Selwyn, J. G.: *Brit. J. Exp. Path.*, **32:** 124 (1957).
160. Sturgeon, P.: *J. Immunol.*, **68:** 277 (1952).
161. Gunson, H. H., and Donahue, W. L., *Vox Sang*, **2:** 320 (1957).
162. Buchanan, D. I., and McIntyre, J.: *Brit. J. Haemat.*, **1:** 304 (1955).
163. Wiener, A. S., and Geiger, J.: *Exp. Med. and Surg.*, **15:** 75 (1957).
164. Geiger, J., and Wiener, A. S.: *Proc. 6th Congress of the International Society of Blood Transfusion*, pp. 36–40, S. Harger, N. Y., 1958.
165. Argall, C. L., Ball, M., and Trentelman, E.: *J. Lab. and Clin. Med.*, **41:** 895 (1953).
166. Rosenfield, R. E., Haber, G., Gibbel, N.: *Proc. 6th Congress of the International Society of Blood Transfusion*. pp. 90–95, S. Harger, New York., 1958.
167. Stratton, F., and Renton, P. H.: *Brit. Med. J.*, **1:** 962 (1954).
168. Greenwalt, T., Sanger, R., and Race, R. R.: cited after Race and Sanger, *Blood Groups in Man*, 2nd ed., 1954.
169. Waller, R. K., and Waller, M., *J. Lab. and Clin. Med.*, **34:** 270 (1949).
170. Allen, F. H.: *10th Ann. Meeting, Amer. Assoc. of Blood Banks*, Chicago, Nov. 4, 1957.
171. Race, R. R.: *Ann. N. Y. Acad. Sci.*, Vol. XLVI, Art. 9, pp. 988–990 (Nov. 8), 1946.
172. Fisher, R. A.: cited by Race, R. R.: *Nature* (Lond.) **153:** 771 (1944).
173. Cappell, D. F.: *Brit. Med. J.*, **2:** 601 (1946).
174. Lawler, S. D., Bertinshaw, D., Sanger, R., and Race, R. R.: *Ann. Eugen.* (London), **15:** 258 (1950).
175. Fisher, R. A., and Race, R. R.: *Nature* (London), **157:** 48 (1946).
176. Fisher, R. A.: *Amer. Scientist*, **35:** 95 (1947).
177. Wiener, A. S.: *J. Forensic Med.*, **2:** 224 (1955).
178. Dodd, B. E.: *Brit. J. Exp. Path.*, **33:** 1 (1952).
179. Stormont, C.: *Amer. Naturalist*, **89:** 105 (1955).
180. Wiener, A. S.: *Bull. Amer. Assoc. Blood Banks*, **9:** 104 (1956).
181. Wiener, A. S., Owen, R. D., Stormont, C., and Wexler, L-B.: *J. Amer. Med. Assoc.* **161:** 233 (1956); *ibid* **164:** 2036 (1957).
182. Coombs, R. R. A., Mourant, A. E., and Race, R. R.: *Lancet* **1:** 264 (1946).
183. Wiener, A. S., and Sonn-Gordon, E. B.: *Rev. d'Hématol.* **2:** 3 (1947).

184. Mourant, A. E.: personal communication.
185. Lewis, M., Chown, B., and Peterson, R. F.: *Amer. J. Phys. Anthrop.* **13:** 323 (1955).
186. Levine, P., Wigod, W., Backer, A. M., and Ponder, R.: *Blood* **7:** 869 (1949).
187. Shapiro, M.: *S. Afr. Med. J.*, **26:** 951 (1952).
188. Wiener, A. S., Brancato, A. J., and Waltman, R.: *J. Lab and Clin. Med.*, **42:** 570 (1953).
189. Allen, F., and Lewis, S. J.: *Vox Sang.* **2:** 81 (1957).
190. Fudenberg, H.: cited by Allen and Lewis.
191. Chown, B., cited by Allen, F.: personal communication.
192. Race, R. R.: *Bull. Centraal. Lab. v. Bloedtransf.* **2:** 191 (1952).
193. Wiener, A. S., Samwick, A. A., Morrison, H., and Cohen, L.: *Exp. Med. and Surg.* **13:** 347 (1955).
194. Wiener, A. S., and Gordon, E. B.: *Amer. J. Clin. Path.* **23:** 429 (1958).
195. Callender, S. T., and Race, R. R.: *Ann. Eugen.* (London) **13:** 102 (1946).
196. Greenwalt, T. J., and Sasaki, T.: *Blood* **12:** 998 (1957).
197. Lawler, S. D.: *Ann. Eugen.* (London) **15:** 255 (1950).
198. Race, R. R., Sanger, R., and Thompson, J. S; cited by Race & Sanger: *Blood Groups in Man*, 2nd ed.
199. Mainwaring, H. R., and Pickles, M. M.: *J. Clin. Path.* **1:** 292 (1948).
200. Mohr, J.: *A Study of Linkage in Man*, 1954, Ejnar Munksgaard, Copenhagen.
201. Mourant, A. E.: *The Distribution of the Human Blood Groups.* 1954, Blackwell Sci. Publ., Oxford.
202. Cutbush, M., and Chanarin, I.: *Nature* (London) **178:** 855 (1956).
203. Cutbush, M., and Mollison, P. L.: *Heredity*, **4:** 383 (1950).
204. van Loghem, J. J., van der Hart, M.: *Nederl. Tijdschr. v. Geneesk.*, **11:** 748 (1950).
205. Wiener, A. S.: *Amer. J. Clin. Path.*, **23:** 987 (1953).
206. Wiener, A. S., and Sonn, E. B.: *Amer. J. Clin. Path.*, **23:** 708 (1953).
207. Ikin, E. W., Mourant, A. E., and Plaut, G.: *Brit. Med. J.* **1:** 584 (1950).
208. Race, R. R., and Sanger, R.: *Heredity*, **6:** 111 (1952).
209. Ikin, E. W., Mourant, A. E., Pettenkofer, H. J., and Blumenthal, G.: *Nature* (London) **168:** 1077 (1951).
210. Allen, F. H., Diamond, I. K., Niedziela, B.: *Nature* (London) **167:** 482 (1951).
211. Rosenfield, R. E., and Vogel, P., Gibbel, N., Ohno, G., and Haber, G.: *Amer. J. Clin. Path.*, **23:** 1222 (1953).
212. Levine, P., Koch, E. A., McGee, R. T., and Hill, G. H.: *Amer. J. Clin. Path.* **24:** 292 (1954).
213. Levine, P., Robinson, E. A., Layrisse, M., Arends, T., and Dominguez Sisco, R.: *Nature* (London) **177:** 40 (1956).
214. Layrisse, M., Arends, T., and Dominguez Sisco, R.: *Acta Med. Venez.* **3:** 132 (1955).
215. Layrisse, M., and Arends, t.: *Science*, **123:** 633 (1956).

216. Lewis, M., Kaita, H., and Chown, B.: *Amer. J. Human Genet.*, **9:** 274 (1957).
217. Layrisse, M., and Arends, T.: *Blood*, **12:** 115 (1957).
218. Wiener, A. S., Unger, L. J., Cohen, L., and Feldman, J.: *Ann. Int. Med.*, **44:** 221 (1956).
219. Shapiro, M.: personal communication.
220. Wiener, A. S., and Brancato, G. J.: *Amer. J. Human Genet.*, **5:** 350 (1953).
221. Davidsohn, I., Stern, K., Strausser, E. R., and Spurrier, W.: *Blood*, **8:** 747 (1953).
222. Hektoen, L., Landsteiner, K., and Wiener, A. S.: *J. Amer. Med. Assoc.*, **108:** 2138 (1937).
223. Davidsohn, I., Levine, P., and Wiener, A. S.: *J. Amer. Med. Assoc.*, **148:** 699 (1952).
224. Wiener, A. S., Wexler, I. B., Stormont, C., and Owen, R.: *J. Amer. Med. Assoc.*, **161:** 233 (1956); **164:** 2036 (1957).
225. Schatkin, S. B.: *Disputed Paternity Proceedings.* 3rd edition, Matthew Bender & Co., N. Y., 1952.
226. Wiener, A. S., Lederer, M., and Polayes, S. H.: *J. Immunol.*, **19:** 259 (1930).
227. Wiener, A. S.: *J. Immunol.*, **24:** 443 (1933).
228. Wiener, A. S.: *Amer. J. Human Genet.*, **2:** 177 (1950).
229. Wiener, A. S.: *The 1950 Proceedings of the International Society of Hematology*, pp. 207–220, Grune & Stratton, New York.
230. Boyd, W. C.: *Amer. J. Human Genet.*, **7:** 229 (1955).
231. Wiener, A. S.: *Ztschr. f. indukt. Abstammungs. u. Vererbungs.*, **59:** 227 (1931).
232. Wiener, A. S.: *Amer. J. Hum. Genet.*, **2:** 177 (1950).
232a. Alvarez, José de Js: Aplicaciones Medico-legales y Antropologicas de los Grupos Sanguineos en la Republica Dominicana. VI Congreso Medico Dominicano, Trujillo, 1951.
233. Wiener, A. S., Owen, R. D., Stormont, C., and Wexler, I. B.: *J. Amer. Med. Assoc.*, **164:** 2036 (1957).
234. Hektoen, L., Landsteiner, K., and Wiener, A. S.: *J. Amer. Med. Assoc.*, **108:** 2138 (1937).
235. Wiener, A. S.: *J. Forensic Med.*, **3:** 139 (1956).
236. Landsteiner, K., Wiener, A. S., and Matson, G. A., *J. Exp. Med.*, **76:** 73 (1942).
237. Wiener, A. S., Sonn, E. B., and Belkin, R. B.: *Proc. Soc. Exp. Biol. and Med.*, **54:** 316 (1943).
238. Wiener, A. S., Zepeda, J. P., Sonn, E. B., and Polivka, H.: *J. Exp. Med.*, **81:** 559 (1945).
239. Wiener, A. S.: *Science*, **96:** 407 (1942).
240. Haldane, J. B. S.: *Ann. Eugen.*, (London) **11:** 383 (1942).
241. Etcheverry, M. A.: *Rev. Soc. Argent. Hematol. Hemater.*, **1:** 166 (1949).
242. Hirszfeld, L., and Hirszfeld, H.: *Lancet*, **2:** 675 (1919).

243. CANDELA, P. B.: *Hum. Biol.*, **14:** 413 (1942).
244. MOURANT, A. E.: *Cold Spr. Harb. Symp. Quant. Biol.*, **15:** 221 (1950).
245. CHALMERS, J. N. M., IKIN, E. W., AND MOURANT, A. E.: *Ann. Eugen.* (Lond.) **17:** 168 (1953).
246. BOYD, W. C.: Blood groups. *Tabul. biol.* (Hague) **17:** 113–240 (1939).
247. WIENER, A. S.: *Blood Groups and Transfusion*, 3rd ed., Thomas, Springfield, 1943.
248. MOURANT, A. E.: *The Distribution of the Human Blood Group*, Blackwell, Oxford, 1954.
249. WIENER, A. S., AND LEFF, I. L.: *Genetics*, **25:** 187–196 (1940).
250. UNGER, L. J.: *J. Amer. Med. Assoc.*, **152:** 1006–1010 (1953).
251. SUSSMAN, L. N., AND SCHATKIN, S. B.: *J. Amer. Med. Assoc.*, **164:** 249–250 (1957).
252. SUSSMAN, L. N., AND MILLER, E. B.: *Rev. d'Hematol.*, **7:** 368–371 (1952).
253. SCHATKIN, S. B., SUSSMAN, L. N., AND YARBROUGH, D. R.: *Crim. Law Rev.*, **2** (1): 44–56 (1956).
254. WYELEGALA, V. B., AND WITEBSKY, E.: *Buffalo Law Rev.*, **7** (2): 209–230 (1958).
255. WITEBSKY, E., KLENDSHOJ, N. C., AND MCNEILL, C.: *Proc. Soc. Exp. Biol. & Med.*, **55:** 167 (1944).
256. WALSH, R. J., KOOPTZOFF, O., LANCASTER, H. Q., AND PRICE, A. V. G.: *Oceania*, **24:** 146–151 (1953).
257. OTTENSOOSER, F., AND PASQUALIN, R.: *Amer. J. Hum. Genet.*, **1:** 141–155 (1949).
258. SIMMONS, R. T., GRAYDON, J. J., SEMPLE, N. M., AND FRY, E. I.: *Amer. J. Phys. Anthrop.*, **13:** 667–690 (1955).
259. SIMMONS, R. T., GRAYDON, J. J., AND SEMPLE, N. M.: *Med. J. Austral.*, **2:** 589–596 (1953).
260. SIMMONS, R. T.: *Anthropos*, **51:** 500–512 (1956).
261. MOLLISON, P. L.: *Blood Transfusion in Clinical Medicine*, 2nd. ed., Blackwell, Oxford, 1956.
262. HARLEY, D.: *Medico-Legal Blood Group Determination*, W. Heinemann, London, 1943.
263. GRADWOHL, R. B. H.: *Clinical Laboratory Methods and Diagnosis*, 5th ed., chapter V, pp. 994–1189, C. V. Mosby, St. Louis, 1956.
264. GONZALES, T. A., VANCE, M., HELPERN, M., AND UMBERGER, C. J.: *Legal Medicine, Pathology, and Toxicology*, 2nd. ed., chapter 27, pp. 634–673, Appleton-Century-Crofts, New York, 1954.
265. COLIN, E. C.: *Elements of Genetics*, 3rd. ed., pp. 108–117, McGraw Hill, New York, 1956.

INDEX